THE
NEW
GENETICS

Challenges for Science, Faith, and Politics

Also by the author:

Forced Options
Wars and Rumors of Wars
Man: The New Humanism
Tangled World
The Existentialist Posture
The Sermon on the Mount

THE NEW GENETICS

Challenges for Science, Faith, and Politics

by Roger Lincoln Shinn

MOYER BELL
Wakefield, Rhode Island & London

Published by Moyer Bell

Copyright © 1996 by Roger Lincoln Shinn

All rights reserved. No part of this publication may be reproduced or transmitted in any form or by any means, electronic or mechanical, including photocopying, recording or any information retrieval system, without permission in writing from Moyer Bell, Kymbolde Way, Wakefield, Rhode Island 02879 or 112 Sydney Road, Muswell Hill, London N10 2RN.

First Edition

**LIBRARY OF CONGRESS
CATALOGING-IN-PUBLICATION DATA**

Shinn, Roger Lincoln.
The new genetics : challenges for science, faith, and politics by Roger Lincoln Shinn

   p. cm.

Includes bibliographical references and index.
1. Challenges for Science, Faith, and Politics. I. Title.
QH447.S55   1996
              95-43419
174′.9574—dc20   CIP

ISBN 1-55921-171-7

Printed in the United States of America.
Distributed in North America by Publishers Group West, P.O. Box 8843, Emeryville, CA 94662, 800-788-3123 (in California 510-658-3453).

To
Ruth Nanda Anshen,
catalyst extraordinaire

CONTENTS

PREFACE | 9
CHAPTER I INTRODUCTION: TWO PURPOSES | 13
CHAPTER II EXCITEMENT AND PERPLEXITY | 17
    The First Great Leap | 19
    The Explosion of Knowledge | 23
    The Human Genome Project and ELSI | 29
    Decisions, Present and Future | 32
    Keeping up to Date | 42
CHAPTER III NATURE, NURTURE, AND FREEDOM | 45
    Nature and Nurture | 46
    Determinism, Chance, and Freedom | 50
    The Oddity of IQ | 55
CHAPTER IV SCIENCE, ETHICS, AND FAITH, IN PUBLIC POLICY | 69
    A Triangular Model | 71
    The Power of Ideology | 85
    The Role of Religious Communities in Public Policy | 89

CHAPTER V   HEALING, ENHANCING, AND
             DISTORTING THE HUMAN | 95
             The Disturbing Question of Norms | 96
             Warnings | 98
             The Interaction of Biology and Culture | 102
             Some Clues to Human Norms | 106

CHAPTER VI   A CRUCIAL ISSUE: MODIFICATION
             OF GERMLINE CELLS | 123
             The Issue and the Spectrum of Opinions | 124
             Scientific Issues in Current Debates | 129
             Issues of Purpose and Values | 135
             Playing God | 143

CHAPTER VII CODA | 147

             REFERENCES | 159
             INDEX | 171

PREFACE

Even more than most books, this one owes its arrival to many people who helped. All writers are debtors to their culture, which gives them an audience to address, a language, a treasury of ideas and symbols, a set of widely-shared purposes and concerns. Even a James Joyce, inventing a unique language, drew deeply on a long heritage of words, memories, and myths. A contemporary interdisciplinary study, like this one, could not have been written a decade ago. And it requires more skills than any individual, certainly than I, can possess.

Here am I, whom nobody would mistake for a geneticist, using contemporary genetics as a case study for examining perennial social and ethical issues. I make no apologies for my amateur standing. In a way it is an advantage. It puts me among the ordinary citizens, their elected representatives, the business people, the foundation executives, the investors who are constantly making decisions in areas where they are inexpert. That is the nature of democracy. But none of us can decide wisely without enlisting the help of many specialists, as I have done here.

Preface

The immediate context of this inquiry is a three-year project of the Center for Theology and the Natural Sciences, in Berkeley, California: "Theological and Ethical Implications of the Human Genome Initiative," supported by a National Institutes of Health grant GNM–1–R01–HG00487–01, Ted Peters principal investigator. The team included specialists in biochemistry, genetics, anthropology, the physical sciences, ethics, philosophy, and theology. And it drew on the immense intellectual resources in the San Francisco Bay area.

Behind that, I owe much to many years of a Columbia University Seminar that met monthly in the home of Ruth Nanda Anshen, chaired first by Margaret Mead, then by Dr. Anshen. The University Seminars are interdisciplinary groups of faculty from various universities and professionals from the wider world. In this one I learned from such eminent geneticists as Theodosius Dobzhansky, Bentley Glass, Francisco Ayala, and Erwin Chargaff. I also entered into conversations on broader issues of science and ethics with physicist I.I. Rabi, poet W.H. Auden, political scientist Hans Morgenthau, mathematicians Richard Courant and Joachim Weyl, sociologist Robert Merton, astrophysicist Fred Hoyle, psychologist Erik Erikson, economist Adolf Lowe, and many others. The influence of that seminar shapes my thought in every chapter of this book. I thank particularly Ruth Anshen, who has an astounding ability to draw diverse and innovative people into conversation, and who stirred me to write this book.

I thank also the American Association for the Advancement of Science and the American Bar Association for their invitation to participate in a national conference on "The Genetic Frontier: Ethics, Law, and Policy," where again my ideas met scrutiny of a wide range of specialists.

More particularly, I thank three people who have read portions of this book and helped me avoid some errors: R. David

Preface

Cole, professor emeritus of molecular biology at the University of California in Berkeley; Ian Barbour, professor emeritus of physics and religious studies at Carleton College and Gifford Lecturer for 1989–90 at the University of Aberdeen; and Audrey R. Chapman, of the Directorate for Science and Policy Programs of the American Association for the Advancement of Science. This might have been a better book if I had not stubbornly resisted some of their suggestions.

Finally, I thank the several colleges and universities that, since 1967, have invited me to lecture on genetics and ethics, and have required me to defend my ideas under the cordial but intensive questioning of faculty and students in all the areas I have touched upon.

I intend this book to be one of many exploratory efforts now underway to meet the critical issues emerging whenever scientific achievements interact with human purposes. Out of the vast range of problems and opportunities, genetic knowledge and power are one tumultuous example.

Roger L. Shinn

Chapter One

INTRODUCTION: TWO PURPOSES

Two purposes govern this book. First, I hope to make a contribution to ethical theory—not simply the theory of scholars, but the theory that operates, often sneakily and sporadically, in every household, every factory and office, every hospital and political process around the world. Second, I want to enter into the public discussions about the uses of the new genetics, specifically the Human Genome Project authorized by the United States Congress and part of an epochal international enterprise.

The two purposes are easy to distinguish. Either alone can become the center of a career. Yet the two are inseparable. Either, in isolation, is misleading and a little ridiculous. An ethical theory, detached from the pressures of decision-making, is abstract and unreal. A plunge into immediate controversies about the ethics of genetics or anything else, with no attention to a more general theory, leads to *ad hoc*, capricious judgments.

Hence the two purposes interact every step of the way. I reject every attempt to start with a theory, then apply it to practice. That method leads to self-deception, to a failure to recognize how the theory has already risen out of the maelstrom of conflicting interests and values in personal and social life. In the richness of human experience people enter into moral struggles and perplexities; there they discover which of their purposes are trivial and which are important; they relate divergent purposes, sometimes choosing among them, sometimes drawing them into an uneasy harmony. Then, reflecting on that experience, they may discover a nascent theory of ethics, raise it to visibility, examine it, smile at its naivete, blush at its crudity, and gasp at its embarrassing self-revelations. They may refine it and purge it of its worst absurdities and inconsistencies. They may compare it with alternative ethical theories of past ages and of the present. In its light, they may revise some of the judgments that entered into its formation. In the constant interaction between activity and theoretical reflection, they may gain some increasing maturity.

This short book points beyond itself to the far greater task of elaborating a comprehensive ethical theory. That has been attempted, never with total success, by a few great minds in human history. Without forgetting that wider vista, I am here focusing on two more particular issues.

In the exploration of ethical theory, I am inquiring into the relation between contemporary science and ethics. I do not underestimate the urgency of many other issues. For example, I fully realize the utmost importance of moral character. Often in human affairs we know our moral responsibility, and the only question is whether we have the integrity and courage to do what is right. But in this era of human history the dramatic achievements of science have forced on the human agenda a set of decisions without any precise precedents. They require exploration as well as fidelity to recognized values. So I here center my attention on the particular problems and opportunities in the relation of science and ethics.

In taking genetics as an example of science, I am again

INTRODUCTION: TWO PURPOSES

sharpening my focus. I do not claim that this example is more important than such others as war and peace in a nuclear age, or ecology, or the daily problems of earning a living, issues with which I have struggled often. I choose it now because it touches on human life so intimately and because it has drawn such public attention. It impinges directly upon our self-awareness. People respond to news reports with a flare of hope or a shudder of fear. The new science evokes language with mystical overtones. Investigators talk of a biological Holy Grail, of "a glorious goal." Books have titles like *The Second Genesis* (Rosenfeld 1969), *The Eighth Day of Creation* (Judson 1977), *The Ultimate Experiment* (Wade 1979), and *The New Genesis* (Cole-Turner 1993).

Even within this area, I am not covering the long, intricate catalogue of issues. I am excluding controversial questions of genetic manipulation in plants and animals; of patents, profits, and the many economic problems in the intimate relating of scientific programs to corporations, governments, and universities; of DNA "fingerprinting" with all its legal and juridical questions; of confidentiality of formerly unknowable information that might be of value to employers, insurance companies, and even prospective spouses. These are immensely important. But here I focus on human genetics as it relates to therapy and possible enhancement of health.

This double narrowing of focus is not a surrender to the contemporary cult of specialization that afflicts research and education today. On the contrary, I hope that by looking closely at the two issues (the relation between science and ethics, and genetic therapies), rather than skimming over many equally important issues, I may throw some light on broad and overwhelming questions of the nature of human life and its relation to all that is.

This restraint means that I am not producing a list of ethical prescriptions. That might be worth doing. Scientific and

Roger Lincoln Shinn

technological projects move like a juggernaut in our world, sometimes transforming life before they can be evaluated. Nuclear fission and fusion are one example. Genetic activity could be another. Ethics must scramble to keep up. So there are times for sounding the alarm and issuing prescriptions—if anybody should happen to listen to them. I want to contribute to some prescriptions, and I do just a little of that here.

But I am impressed with Alfred North Whitehead's insight: "Every philosophy is tinged with the colouring of some secretive imaginative background, which never emerges explicitly into its trains of reasoning" (Whitehead 1925, 11). Before approving or rejecting prescriptions, we can look for their rootings in that "secretive imaginative background." That sometimes means building some speed bumps on the road that prescribers travel. It means challenging all the contestants in furious ethical controversies, "How do you get to that conclusion? How do you know you are right? What might possibly require you to change your mind?"

I quickly grant that society needs polemicists. When they are right, they create new possibilities. Even when they are wrong, they often jar a public into awareness. Without past polemicists against established authority, there would be no modern democracy. To refrain from action is a kind of action. It is, in effect, an aquiescence in the drift of events, whether that be an impersonal process or the manipulation of people with their own purposes. So polemics and crusades belong in social life.

However, this book is not a polemic or a crusade. It is intended as an inquiry that leads to some judgments and raises some questions that require public attention. I hope that it is something between a tract for the times and a contribution that will continue to have some value a little beyond the day after tomorrow.

Chapter Two

EXCITEMENT AND PERPLEXITY

Heredity has an age-old fascination, rational or absurd, for people everywhere. As far back as we can look in history, families and tribes have felt the bonds of kinship. For better and for worse, mortals, destined to die, have sought to perpetuate themselves in their children and their children's children. They have rejoiced in family resemblances. Kings have tried to immortalize their power by establishing dynasties, and peoples have accepted the dynasties on the dubious assumption that heirs would have some of the charisma of a royal ancestor. Societies have institutionalized their beliefs and prejudices about heredity in systems of class, caste, and slavery.

Men, preoccupied with heredity, have foisted injustices on woman. At the birth of a baby, the mother is present and recognizable. The father may or may not be identifiable. So men, possessive about genuine descendants, have imposed sexual disciplines on women that the men had no intention of accepting for themselves. This is one of the causes of the double standard of much sexual morality, whether in the crude form of medieval chastity belts or in more genteel behavioral codes.

Beliefs about heredity through the centuries have been mostly nonsense. They are imbedded in epics and sagas, myths and legends, folk tales, and behavioral customs around the world. Anyone who reads the Bible as a book of science will be jolted by some of its stories. Notice how Jacob cheated his father-in-law by a ridiculous method of managing the breeding of sheep (Gen. 30:37–43), a trick that Shakespeare remembered in *The Merchant of Venice*. Plato, that eminent intellectual, would have laughed at Jacob, but his own ideas about heredity were not much better. Typically heredity has been associated with blood, as in talk of "blood relatives," or of "blue blood" to denote aristocracy. "Blood is thicker than water," goes the old saying. Although unscientific, the language of blood is a symbolic recognition of the actual power of blood in the human body.

In spite of their ignorance about heredity, most cultures developed bits of practical wisdom out of experience. Tabus against incest are an example. Farmers learned to propagate seed from their best plants. Herders chose their best bulls and horses—that is, the best for human purposes—to sire offspring. In Darwinian terms, they were modifying natural selection with a deliberate human selection. The choice of spouses, whether in arranged marriages or voluntary marriages, used many criteria, including the desirability of a prospective spouse as a parent. That desirability was far more than a matter of biological heredity, but ideas of heredity, whether true or false, entered into the decisions.

Through most of the millennia of human history, men and women groped for knowledge about heredity in a confusion of superstition, wishful and fearful experience, and everyday empirical observation. Today we are still trying to penetrate the age-old veil of ignorance. But scientific discoveries, emerging in the nineteenth century and exploding in the twentieth, have brought radically new promises and perils to our human race.

This chapter offers a glimpse of the present scene. It is far from a summary of modern knowledge. Others have done that with professional skill. I want only to record a little of the story of discovery, in order to catch a bit of its excitement, and to point out some of the new powers that require ethical decisions.

The First Great Leap

It was late in history that the first great leap in genetic science occurred. Such giants as Copernicus, Galileo, and Newton had already transformed human understandings of the cosmos, when an obscure Austrian monk, Gregor Mendel, made his revolutionary discoveries in genetics. His superb scientific achievement combined gardening, the youthful science of statistics, a monastic-style diligence, and flashes of sheer innovative insight.

His chief experiments, as many high school students learn, were with peas. He became curious about seven traits in peas—size, wrinkled or smooth surfaces, color of seeds, and some more. Here I take color as the example. Mendel cross-pollinated green and yellow peas. Common sense might expect that the result would be a mixture of some yellow and some green peas or a yellow-green blend. Instead, all the peas were yellow. Common sense then might ask, what happened to the green? Did it just get lost?

Mendel asked with a scientific curiosity. He cross-pollinated the first generation hybrids (to use a later vocabulary), and the green reappeared. Most of this new generation were yellow, outnumbering the green by about 3:1. But why, Mendel asked, would two yellow parents produce green offspring? He continued his experiments, growing thousands of plants, inspecting them, counting, and recording the results. And he formulated the Mendelian laws that are still important for genetics.

Roger Lincoln Shinn

The word gene had not yet been invented. For a monk, Latin was a more familiar language than Greek, and Mendel used the terms *elements* and *characters* to describe the hidden force he was discovering. He even used the term *dominant*, and later scientists added the term *recessive*. In Mendel's peas, yellow was a dominant, green a recessive characteristic.

In the first generation of hybrids, each pea had one gene-for-color from each parent. That meant a gene-for-yellow and a gene-for-green, and the gene-for-yellow won out or dominated. But the recessive gene-for-green was still there. We might represent this generation as:

Yg gY Yg gY Yg gY Yg gY Yg gY

Here the **Y** for yellow is a capital letter because yellow is dominant; the **g** for green is lower case because **g** is recessive.

When Mendel then cross-pollinated peas from this generation, there were four possibilities. The pea in the next generation might have a **y** from both Mom and Pop; it might have a **Y** from Mom and a **g** from Pop; it might have a **g** from Mom and a **Y** from Pop; it might have a **g** from both Mom and Pop. So the four possibilities are:

YY Yg gY gg

That means there are three chances for a yellow pea and one chance for a green pea. Looking at those three yellows, a gardener or cook would not know that one (the **YY**) is as pure-bred a yellow as its grandparents, while two (the **Yg** and **gY**) have a recessive gene-for-green in their innards, which they can transmit to their children. Similarly, the green pea is as purely green as its grandparents, and it is not the least bit bothered that green (in relation to yellow) is recessive, because it does not know that and does not care. It can be entirely comfortable in its unadulterated greenness.

Once again, a later vocabulary helps to express what Mendel discovered: the difference between genotype and phenotype.

Excitement and Perplexity

The **YY** and **Yg** peas look alike, but only the **Yg** has that recessive gene-for-green that might emerge in some later generation. So we can say that the phenotypes (the observable peas) are the same, but the genotypes (the hidden genetic structures) are different. That distinction later became very important for human genetics.

We should note in passing that Mendel's discoveries deal with statistical probabilities. Reality rarely corresponds exactly with such probabilities. For example, in flipping coins the statistical probability between heads and tails is 50–50. That does not mean that in a hundred tries 50 flips will result in heads and 50 in tails. We should be surprised at any such exact results, but not as surprised as if we got 95 heads and five tails; in that case we should suspect a "fixed" coin. Over the long run the reality tends to come close to the statistical probability.

We should notice one other warning about statistical probability. In flipping coins, we do not expect to get five heads in a row. That, though improbable, is not impossible. If we do get the five in a row, that tells us nothing about the sixth flip, despite some gamblers' hunches that either a hot streak will continue or that it's time for tails to come up. On the sixth flip, the chance of heads is 50–50, the same as on the first flip. The coin does not remember what it has been doing or decide that it's time to straighten out the odds.

To talk of flipping coins may seem a silly digression from a subject so profoundly serious as human nature and its inheritance. But one thing we learn from genetics is a confirmation of an ancient insight. This human personality, exalted though it may be, is imbedded in the most common and elemental processes of nature. As the Book of Genesis puts it, the human creature is made of the dust of the earth. That a divine spirit may infuse that dust with the breath of life is a reminder of other dimensions to selfhood. But these do not annul the primal understanding expressed in phrases like "ashes to ashes, dust to dust."

It has been alleged that Mendel "cooked" his figures. That is, once he made his discovery, he may have skewed the data a little toward the statistical norm, to make them convincing to others. Such fudging of data to fit a theory occurs every now and then in reports of scientific experiments. Today, if discovered, it creates a scandal. If Mendel did it, we can say that a scientist, who was also a monk, should have been strictly accurate and honest. But at least he was not doing it to win a corporation contract or a Nobel prize.

Mendel's work, for all its originality and brilliance, was basically simple. It took no elaborate equipment, no lavish grants from government or foundations, only the indulgence of his fellow monks. Yet biomedicine keeps returning to his work, both as an example of scientific method and as a useful contribution to practice. Today some lethal diseases follow a Mendelian pattern. Some, though not most genetically-transmitted diseases are monogenic; that is, a single gene causes the disease. And in some cases both parents are phenotypically free of the ailment, but they have the recessive trait—like some of Mendel's yellow peas that had the recessive gene for green. If they conceive a child, there is a one-in-four statistical probability that the child will have the ailment. I'll come to some examples soon.

Perhaps Mendel deserved the earth-rattling fame of a Galileo. But he did not get it. He wrote a paper, read it at a meeting in 1865, published it in an obscure journal the next year (Mendel 1866). Nobody paid much attention. His discoveries were, for all practical purposes, lost for decades. Although Mendel knew Darwin's *Origin of Species* (1859), Darwin never learned of Mendel. Darwin, who acknowledged that he knew little about heredity, might have saved himself some big mistakes if he had known of Mendel.

Maybe the lack of fame did not bother Mendel. He became abbot of his monastery in 1868 and was too busy for scientific experimentation. He died, unsung, in 1884. His work came alive in the twentieth century.

The Explosion of Knowledge

Toward the end of the nineteenth century some excellent work in biology prepared the way for bigger things to come. At the turn of the century three scientists, independently, rediscovered Mendel. Then came a rush of activity, combining painstaking laboratory experiments and occasional revolutionary discoveries. The whole story is far too complicated for this short book. What I can do here is tell enough of it to point to the issues that require ethical decisions.

The human person, we keep rediscovering, is a marvelous organism. Each of us is fragile and intricate, embodying many processes that can go wrong in so delicate a body. But the whole organism is awe-inspiring.

The adult body, by some estimates, includes 100 trillion cells, give or take a few trillion (Lee 1991, 4). That figure is incomprehensible even to those of us who worry about the national debt. My electronic calculator cannot handle numbers that big, and nobody can count that high. But my body, when I am healthy, coordinates all those cells in a living organism. A cell is tiny, but is visible in a microscope. Its complexity strains imagination. Biologist Lewis Thomas writes: "My cells . . . are ecosystems more complex than Jamaica Bay" (Thomas 1974, 2).

Within most of these cells—the red corpuscles in the blood-stream are an exception—is a nucleus. Each nucleus includes twenty-three pairs of chromosomes—except for the germ cells (sperm and eggs) that have twenty-three chromosomes, not twenty-three pairs. (The forty-six chromosomes are nothing to boast about; a turkey has eighty-two.) Each pair includes one chromosome from the individual's mother and a matching (homologous) one from the father. That's a simplification; because of "crossing over," a chromosome itself may come partly from the mother and partly from the father. Thus siblings share some chromosomes and differ in others. As everybody has noticed, siblings are likely to have remarkable resemblances to each other and to their parents, and equally remarkable differences. Only identical twins, not all twins, have exactly the same genotype. A sister and brother cannot be identical twins. Two sisters might or might not be identical twins; the same for two brothers. The many possible combinations of patterns of chromosomes account for the many different heredities among people in the same family.

There is one other cause of variation: mutations. In the intricate processes of genetic combinations, sometimes there is a spontaneous, or accidental, variation within a chromosome, causing a change in the pattern of heredity. Major departures of organisms from the expected hereditary patterns are likely to be such mutations. Gardeners call them sports. Most new varieties of plants are the result of cross-pollination, a favorite activity of bees and some people. Hybridizers, working in the tradition of Mendel, produce most of the tempting new plants featured in each year's seed catalogues. But an occasional new variation is a sport that may be noticed by a sharp observer and weeded out or selected for propagation.

The chromosomes within the nucleus of each cell are amazing combinations of chemical materials and processes. A chromosome is a macromolecule. That is a very complex and large molecule (large for molecules but tiny compared with the objects we see with our senses). Each chromosome contains a tightly-coiled strand of deoxyribonucleic acid. Since that is hard to pronounce, it is called DNA for short. Stretched out in a line, the

Excitement and Perplexity

forty-six human chromosomes would extend more than six feet. If the six-foot length in each of 100 trillion human cells could be stretched out straight, that would be 600 trillion feet or well over 113 billion miles. That, incredible as it may seem, would reach to the sun and back 610 times with a lot of miles to spare. If it seems impossible that all that stuff should exist in one human body, we can recall that those chromosomes are extremely skinny—their length is some hundred million times their width—and tightly coiled.

Herman Muller, a Nobel laureate in biology, estimated in 1933 that the total quantity of human germ plasm amounted to scarcely one teaspoonful for the entire world (*New York Times*, April 26, 1955, 17). Allowing for the population explosion, we might reckon two or three teaspoonfuls by now. The perpetuation of the human race and its hereditary legacy depends on those teaspoonfuls of germ plasm—distributed among all the people of the world. It is no wonder that some geneticists think of germ plasm with near reverence. Erwin Chargaff, who contributed so much to the discovery of the composition of DNA, wrote of the "lyrical shudder" that moves the scientist "to tears" before the mysteries of nature that we can "only touch tangentially" (Chargaff 1978, 112).

The chromosome is a string of genes, tiny packets of DNA. If we ask how many there are in the human genome, nobody knows. I have seen recent estimates ranging from 50,000 to 240,000, but they are tending to converge at about 100,000. It may seem strange that, given the many discoveries of recent years, nobody has yet counted the genes. But they do not have clear signposts marking their boundaries. The gene is better understood functionally than spatially. It is the bit of DNA that carries the code that, via an intermediate process, produces a specific bodily protein. In the strange ways of nature, there can even be genes within genes. Some

Roger Lincoln Shinn

geneticists say that we may never get a precisely accurate count of the number of genes.

In the 1940s, scientists began to discover the composition of DNA (and, therefore, of chromosomes and genes). DNA is made up of tiny chemical bits called nucleotides. Each nucleotide combines a nitrogen-containing molecule, a five-carbon sugar, and a phosphate group. There is nothing especially remarkable about these chemicals, as such, but the combination in DNA has an amazing potency. An average chromosome—they vary in length—includes about 140 million pairs of nucleotides, but they all fall into four types, labeled **A,G,C,T**. (Those are the abbreviations for adenine, thymine, guanine, and cytosin.) The four are often called the alphabet of heredity. All of us have this same four-letter alphabet in every cell nucleus of our bodies.

At first glance it may seem incredible that all the varieties of human personality around the globe and through thousands of years of history, to the extent that they are hereditary, can be spelled out with a mere four-letter alphabet. But in digital communication everything (words, pictures, and music) is "spelled" with combinations of ones and zeroes. So four letters are enough to encourage a variety so great that it constantly astonishes us. Computer scientists have actually used DNA—nature's digital bits—to solver computer problems, and some think that future DNA supercomputers will have abilities far beyond the best current silicon capacities.

We human beings share most of this genetic alphabet with other forms of life. "Our chromosomes differ from those of the chimpanzees in approximately 1% of the genes. . . . Precious little DNA differs between the zoo keepers and the zoo inhabitants" (Lee 1991, 13). It is possible to combine genes from bacteria, yeasts, plants, and animals. Scientists study the mouse genome because most human genes can be found in a mouse. They are transplanting human genes into a mouse in order to study Alzheimer's disease. In many ways we have a shared identity with all forms of life—a blow to human arrogance. "Ants are so much like human beings as to be an embarrassment," writes Lewis Thomas (1974, 12). But the distinction remains immense. One small example is that only human beings—not bacteria, ants, mice, or chimpanzees—are now learning the composition of DNA and the ways of turning that knowledge into vast new powers.

The second half of the twentieth century brought two great discoveries, so spectacular that they were quickly spread over every tabloid and TV tube. The first was the discovery of the structure of DNA—the now-famous double helix. In 1953 two young scientists, the American James Watson and the British Francis Crick, made the decisive breakthrough in the Cavendish laboratory in Cambridge, England. They have each told their stories in short books that reveal the grand excitement, the competitive rivalries, the high jinks, the low comedy, and the ecstasy of science—quite different from the prosaic discipline that people sometimes attribute to scientific research. The contrasts between the memories of Watson and Crick, and also the memories of some others, tell something about human quirkiness, but they take nothing from the significance of the achievement (Watson 1968; Crick 1988; Chargaff 1968; 1974; 1978, 100–102).

The next great breakthrough, in 1973, led to recombinant DNA. That is, scientists began to learn to slice up DNA, rearrange the sequence of genes, and even substitute one gene for another. In 1967 Herman Muller had contemplated such acts—surgery by nano-needles, as he called it—and declared them impossible (Muller 1968, 255). He saw no conceivable way of severing genes from their neighbor genes. As late as 1970 Jacques Monod, another Nobel laureate, said the same thing even more emphatically (Monod 1970, 164). And then it happened. Chemicals can do

Roger Lincoln Shinn

what knives cannot do. Some enzymes, known as restrictive enzymes, act like chemical scissors. Even more remarkable, sometimes the "right" enzyme will head for the "right" place in the strand of DNA and sever the "right"—that is, the desired—gene. Then another fragment of DNA can be inserted into the resulting gap. The delivery agent, called a vector, is usually a retrovirus (a genetically-modified virus). Thus it becomes possible to rearrange or recombine DNA. The scientific term is *recombinant DNA*; the popular term is *gene-splicing*. Some of this occurs naturally, as bits of "rebel DNA" move around in a chromosome. Now scientists can do it purposefully.

Suddenly the way opened to a fervor of activity. Here was the utterly new possibility of intervening in the genetic composition of bacteria, or vegetables, or animals—and, with more trepidation, of human beings—in order to initiate changes undreamed of in all past history. There were also possibilities of a disastrous mess, if restrictive enzymes cut DNA or vectors inserted DNA in the wrong places, or if the new bit of DNA did not get properly integrated with the host DNA.

Even experiments on bacteria had their dangers. Might experimenters unintentionally produce a new lethal bacterium, to which the human body had none of the resistance that has been built up, over the centuries, to other perils? Scientists were themselves so concerned about the dangers that in 1974–75, at a meeting in Asilomar, California, they adopted a self-imposed moratorium on experimentation. Then the National Institutes of Health in the United States issued a code for experimentation. As might be expected, the code drew criticism. Some thought it too tight, inhibiting valuable research. Others thought it too loose, still permitting dangerous experiments. Further experience has brought relaxation of the code, with controversy every step of the way.

Excitement and Perplexity

The Human Genome Project and ELSI

As geneticists began to locate specific genes with specific consequences, they conceived the project of mapping the human genome, i.e. the complete array of genes in a person's cells. The proposal was ambitious and expensive, with an estimated cost of $3 billion over some fifteen years. The U.S. Congress made an initial appropriation of $135 million, and the project was formally inaugurated in 1991, with more appropriations following. The aim is to locate every gene in human DNA. The work goes on in many universities and research centers. In utter contrast to the original development of nuclear weapons, the international cooperation, helped by the privately-funded Human Genome Organization (HUGO), is unprecedented.

Beyond the "mapping" of the genome is the more detailed "sequencing." That means locating the nucleotides (the **A,G,C,T** alphabet) that compose the genes. If accomplished, it would amount to perhaps three billion letters. A printout, some reckon, might require a stack of books the equivalent of thirteen sets of the *Encyclopedia Britannica*. Maybe there would be no reason to print it out; nobody wants to spend many evenings reading variations on **A,G,C,T,** and the information would be more accessible in computers. Also, because of individual differences, amounting to about one out of a thousand bases, there could be no single map or sequence identical for everybody. If people differ from chimpanzees in about 1 percent of their genetic material, individuals differ from one another in about .1 percent. Hence a representative map would have most of the information about everybody.

The Human Genome Project (HGP) shows an important aspect of science as a human and social activity. When the Nobel prize was awarded to Watson and Crick (along with the London biologist Maurice Wilkins) for the discovery of the double helix, "a spokesperson for the awarding institution noted that their contribution had 'no immediate practical application'" (Shapiro 1991, 76). But when Watson, about 25 years later, urged a Congressional committee to appropriate money for the HGP, he wanted "to mobilize the country to do something about diseases" (Watson 1968, 168). Why the difference? Partly because those who do "pure science" for the sheer love of discovery never know what consequences may follow from new knowledge. Partly because gene-splicing, added to the double-helix, transforms esoteric knowledge into a powerful agent of change. Then society—yes, the whole nation and the world of nations—has a stake in the enterprise.

For this reason Watson, who became the first director of HGP, immediately emphasized the ethical issues involved in the project. Three percent of the total budget was allotted to research on the *ethical, legal, and social implications* (ELSI). A national committee oversees the research, and it goes on among individuals and interdisciplinary teams in many parts of the country.

I must immediately state my qualified endorsement of ELSI. I agree, without qualification, that the ethical issues in HGP are inescapable, urgent, and utterly important. Without sharp attention to them, we may find ourselves doing things (as we did with nuclear weapons and energy) before we realize what we are doing. Money is one of the ways our society uses to focus attention on issues. But money has its problems.

The first qualification on my enthusiasm is a warning against false expectations. ELSI is not a vending machine that delivers a product when money is inserted. Neither, of course, is scientific research. The flashes of insight that bring scientific revolutions are neither predictable nor purchasable. But at a certain stage of science—the development and elaboration in detail of a new insight—it is more or less the case that you get what you pay for (Kuhn 1970). HGP started from works of scientific genius. It proceeds, in somewhat predictable ways, by the work of skilled technicians and computers. At any stage of the work, it is possible to say, "In the past year (or month) we have accomplished this, this, and this. In the next stage, we expect to go on to this, and this. Our results thus far are verifiable in laboratories anywhere in the world. Others can accept our findings and go on to make their own." Thus different universities and research centers can divide up the job and coordinate their efforts. The mapping is actually ahead of schedule. A periodical publishes charts on twenty-some research centers, showing the goals of each and the accomplishments thus far (*Human Genome News*; e.g., Vol. 6, No. 4, Nov. 1994).

Ethics is not like that. Neither is politics or art. A musician cannot say, "Because Mozart did thus and so, I can regard that as an accomplished job and push on to something more adequate. Or, because Columbia University is working on opera, MIT can work on symphonies, and Cal Tech can work on lieder, and a coordinating committee can put them all together year by year."

Ethicists are wiser for knowing the insights of religious prophets and saints and the scholarly works of Plato, Aristotle, Augustine, Spinoza, Kant, and many others. But those past achievements are all still in controversy. And when it comes to the new ethical problems, nobody can say, "Because Harvard is working on risk analysis, the University of Chicago should work on criteria of health, and Oxford on informed consent."

So ethical inquiry is likely to appear muddled, as compared with laboratory research. But so are statecraft, marriage, artistic creation, and most of what is important in life. To see the muddles is not to discredit the activities; it is to understand something about the human condition. So I endorse ELSI, and expect good of it, but I warn against glib expectations.

The second qualification is that money influences processes of research. To begin with, it affects the direction of work.

Roger Lincoln Shinn

Critics of HGP sometimes complain that it lures scientists into funded genetic research when they might better be working on—name your favorite project—AIDS, the recurrence of tuberculosis, infant mortality, or whatever. Such criticisms are legitimately debatable. Similarly, ELSI may attract specialists (biologists, legal scholars, social scientists, ethicists) who might better be working on the desperation of the poor, the instability of families, or domestic violence and international war. Such questions, too, are legitimately debatable.

A more insidious charge is that ELSI is an attempt, subtle or not so subtle, to co-opt potential critics into support of HGP. A person or a team, if funded by ELSI and perhaps hopeful of a renewal of a grant, is not likely to come to the conclusion that HGP is a monstrous mistake. In saying this, I am not charging that HGP is a mistake; I am only saying that genuine ethical research should not rule out the possibility that HGP is misdirected. The Council for Responsible Genetics (based in Cambridge, Massachusetts) is skeptical, maybe even cynical, about the whole project.

In the interest of "full disclosure" I therefore confess: Critics, be warned. My opinions may be tainted. I have participated in transdisciplinary study projects funded by ELSI. Federal money, minute by criteria of the total HGP but useful to me, has funded travel and research. I have learned immensely from these experiences. I am also a member of the Advisory Board of the Council for Responsible Genetics. I do not think that these diverse, sometimes contradictory associations have corrupted my opinions and commitments. But that is for others to judge. With that warning, I return to my present task.

DECISIONS, PRESENT AND FUTURE

The new knowledge and new possibilities of action require new decisions. These fall into several categories.

Excitement and Perplexity

1. **Diagnosis**

Some genetically-transmitted ailments are monogenic, carried by a single gene. For reasons that Mendel would have understood, a disastrous disease may suddenly erupt in a person whose parents showed no sign of it. Now it is possible to diagnose individuals to discover the presence or absence of some portentous genes. One of the earliest effective diagnoses was of Tay-Sachs disease. More recent discoveries include Huntington's disease, muscular dystrophy, and cystic fibrosis.

More often the story is far more complicated. Some ailments are polygenic, with no single gene that we can "blame." In other causes, genes produce only a susceptibility to a disease that has other causes. Week by week scientists are making new discoveries. Recent work has shown some genetic susceptibility to various types of cancer, multiple sclerosis, diabetes, perhaps alcoholism and obesity, and some types of the very complex Alzheimer's disease. Here, for the sake of simplicity, I shall deal mainly with monogenic problems.

But first I must make an important digression. Some ailments affect specific ethnic populations. Cystic fibrosis, the most frequent hereditary disease in the United States, is found mainly in people of West European descent, among whom about one in twenty carries the recessive gene and about one in 2,000 is born with the disease. Tay-Sachs is most common among Ashkenazi Jews (that is, Jews of East European descent, roughly the Yiddish culture). Within that population about one in 3,600 children is affected, with a far higher percentage when parents are first cousins. Sickle-cell anemia is most frequent among people of African and Mediterranean descent, where the recessive gene gives some resistance to malaria, but a double dose (from both parents) produces the serious disease; about 8% of American black people have

[33]

the recessive gene, and about 2% have the disease. Such associations feed racial prejudices in a society already inflamed with prejudice. The truth is that all groups have their vulnerabilities. Rather than stigmatize groups for their special problems, we can better work together to find ways of preventing, healing, or living with the perils that are part of our fragile human nature.

Diagnosis is a benefit when a treatment is available. For many years, doctors have tested newborn infants for *phenylketonuria* (PKU), a genetically-based ailment. In most states the test is compulsory—for the good reason that immediate discovery can lead to a diet that will prevent mental retardation. With some other ailments a prenatal diagnosis is possible, with the possibility of treatment in utero or immediately after birth. In an increasing number of cases, genetic testing can be carried out on a fetus before it is born or on parents before they even conceive a child.

Diagnosis is not treatment. Often it leads to treatment, but the delay may be many, many years. Controversies rage about the value of diagnosis when no treatment available. For a time some states proposed compulsory screening of black people for sickle-cell anemia, even though there was no known treatment. The idea was soon dropped, because it put a stigma on blackness without a benefit. Recent evidences of an effective treatment for sickle-cell have led a federal panel and several professional panels to urge screening of all newborn infants.

One case, important in itself and possibly a harbinger of other things to come, is Tay-Sachs disease. This devastating disease causes the degeneration of the nervous system and leads to death, agonizing to child and parents, usually by the age of four. There is at present no effective medical treatment. The test for prospective parents can show whether they carry the gene. Since it is recessive, it will not harm the child unless both parents carry it. In that case, their child has a one-in-four possibility—the familiar

Mendelian pattern—of suffering the disease. Diagnosis of prospective parents, at their own request, has become frequent in populations at risk.

    The painful trouble is that there is no known effective treatment. Often an engaged couple asks for diagnosis. If it shows that both carry the gene, they face a hard choice. They may reluctantly break their engagement and seek other partners. They may marry with the resolve to remain childless or resort to AID (artificial insemination by donor), after assurance that the donor is free from the Tay-Sachs gene. They may conceive a child, knowing that the chances are three to one that the child will not suffer Tay-Sachs. They may, and sometimes do, conceive, have the fetus tested in utero, and resort to abortion if the fetus is afflicted. Then they may want to try again with a three-to-one probability in their favor. They cannot figure that they have used up their bad luck and can expect good luck next time. (Remember my earlier comments on statistical probability.) The prevalence of Tay-Sachs dropped by about 80 percent in one decade (1970–80), primarily because of voluntary screening and restraint from marriage between carriers.

    Not all genetically-transmitted ailments follow the one-in-four statistical probability of Tay-Sachs. Huntington's disease, which brings a degeneration of nerve cells in the brain with consequent death, results from a dominant gene. Because the symptoms usually appear in mid-life, a parent may conceive children with no knowledge that those children have a one-in-two risk of inheriting the disease. Nancy Wexler, whose mother died from Huntington's disease, led a gallant and skillful study of the ailment. Only after a long search has a test been discovered. There is still no known treatment. Nancy Wexler herself went on to become head of the national ELSI project.

    A final comment on diagnosis is important. There is no possible guarantee now—probably forever—that a fetus will become the "normal child" that parents crave. The most that is possible is a scientific assurance that the fetus is free from some specifiable testable ailments. But all of us carry a few undesirable recessive genes and a few genes that restrict or harm our health. Usually we learn to live with them. We can do much to improve our health, whether by genetic acts or healthful living. But we remain vulnerable beings, subject to illness and inevitably death. No ethic has a right to forget that. The religions that underlie most ethics frequently remind us of our own frailty and finitude.

2. Abortion

Abortion is a subject distinct from diagnosis, and it requires separate discussion. Most abortions have nothing to do with genetics, and the controversies over abortion rage furiously in American society. I cannot here go into the larger issues, but I shall assume that the Supreme Court decision in *Roe vs. Wade*, though not the last word on the morality of abortion, sets a context for discussion of public policy in this country, at least for the time being.

In a society that permits abortion for any reason at all, the knowledge of a truly severe genetic defect is one of the strongest of reasons. The woman or the couple involved might believe it better to prevent the conception (if possible) of a fetus doomed to radical disability. Even so, they might choose abortion when they confront the fact of an ailing fetus. But one problem remains. Amniocentesis is the most frequent method of prenatal diagnosis, and the results of amniocentesis are usually not available until the fourteenth to sixteenth week of fetal development—well after the first trimester, which is so important in present juridical law. Chorionic villus sampling can take place earlier, in the ninth to twelfth week; but there is still some controversy over the risks involved. For some purposes, still earlier methods of diagnosis, ultrasound or procedures now unknown, may push the time still earlier.

The most frequent reason for prenatal diagnosis in this society is Down syndrome. It was formerly called mongolism—evidence of the creeping of racist ideology into purported science. It is not, strictly speaking, inherited. It is a chromosomal disorder, more frequent among older than younger mothers (and possibly fathers, although this is not known for sure). It is not nearly as severe as some other disorders, but it is a major disappointment to parents. It raises the issue: where, if anywhere, on the spectrum from totally disabling diseases to the mild disorders that annoy all of us, do we draw the line that destroys a nascent person? That question will not leave us soon.

One example is poignant. It appears almost certain that in some countries—particularly the two most populous countries in the world, China and India—the most common reason for "selective" abortion is that the fetus is female. It happens less frequently in the United States. Many people, although they accept abortion, readily or reluctantly, find it morally offensive to define female gender as a "disease" justifying diagnosis and abortion. The Chinese government has enacted laws, effective in January, 1995, to prohibit sex-screening of fetuses. In this country, some doctors, refuse to abort fetuses for reasons of gender; some, performing a prenatal diagnosis, refuse to disclose the sex of the fetus. But there are always others less inhibited.

The question becomes urgent: what do we define as normatively human, and what do we define as disease? That will be the subject of Chapter 5.

3. Genetically-engineered medicines

One application of the techniques of gene-splicing is the production of medicines. An example is insulin, normally made by the pancreas. For years it has been extracted from animals and used by humans for the treatment of diabetes. The new process inserts human genes for insulin into bacteria, which then multiply, producing human insulin, which is better, at least for some purposes, than animal insulin.

Similarly, interferon is a genetically-engineered protein that increases resistance to viruses. Earlier enthusiasm for interferon exaggerated its efficacy and overlooked the risk of side effects, but it still confers benefits.

The value of such drugs is real, provided they are used with due caution about risks, a caution that is necessary with all new drugs. But there are controversies. One currently centers on a hormone—it can be extracted from slaughtered cows, but is cheaper when genetically engineered—that increases milk production among dairy cows. Advocates for animals point out that the drug does not improve the health of cows; at best it makes them more useful to human beings. Some critics argue that the hormone threatens the health of cows and of people who drink their milk; others argue against its economic benefits. But the use of the treatment is on the rise.

Other controversies center on human growth hormone, which can increase the height of some short children, developed as a remedy for dwarfism. In some ways it is not a radical treatment; it corrects the deficiency of natural growth hormone, produced by the pituitary gland. Increasingly, parents are asking for it for their children who, by medical standards, are not "abnormally" short. There are many arguments about its actual efficacy. If these are settled, other arguments arise. The frequent example concerns the basketball player who never made it in the big time because he was only 6'4". Has he a right to give growth hormone to his son—or daughter—hoping to add a few inches to the child's height?

Suppose, then, the child prefers violin playing to basketball and has to cope with this awkward height throughout life.

Or consider the fact that most people, it seems, would prefer to be a little taller than average. It is an absolute impossibility—unless in Garrison Keillor's Lake Wobegon—for everybody to be above average. Will we see a continuing race for drug-induced height, which can only end in a mix of winners and losers about the same as before the race began? The cost of the treatment is estimated at $20,000 to $40,000 per year, and drug manufacturers aggressively promoted their product, until public complaints stopped them. Since most families cannot afford that for one child, let alone two or three, will access be determined by the more or less political processes of Medicaid or by the private decisions of insurance companies? This issue will come up again and again as costly new genetic processes develop.

4. Direct somatic therapy

The next step is far more radical. Instead of genetic engineering of medicines to treat people, it is the direct intervention in the human body to modify the genetic structure within the cells of the body.

To repeat a point, everybody has a few genetic liabilities. If mine is astigmatism, I'd far rather wear glasses than have somebody tinker with my genes, hoping to improve my vision without glasses and possibly damaging it. If my problem is diabetes, I'd rather, at least in the present stage of knowledge, take insulin (perhaps genetically engineered) than invite a repair of my genes. But if I had Huntington's disease, for which there is no known treatment, and if there were a possibility—presently as improbable as heart transplants a few years ago—of correcting the defective gene, I might welcome the effort. If told that the attempt is risky,

Roger Lincoln Shinn

I might accept that risk as better than the 100% risk in doing nothing.

The work on plants is already going ahead. The Flavr Savr tomato is on the market amid much publicity. Corn, cotton, and potatoes have been genetically altered so that they make their own pesticides. The move from vegetables to people is a tremendous leap. With plants, the genetic technician, like the hybridizer, can accept a thousand failures for the sake of one success. With people, something called human dignity or the sacredness of personality inhibits rash experiments, even careful but risky experiments.

The complexity of the human genome is part of the problem. A single human gene may encode ten or twenty different functions in different tissues. The effort to correct one function could conceivably derange nineteen others. Again, most ailments—one estimate is 98%—are polygenic. It is reckoned that sixty-three genes determine the color of a mouse. Modification of three or four might fail to bring about a desired change, while introducing other liabilities.

Still, there may be hopeful therapies, particularly in the case of monogenic liabilities. Possible treatments involve inserting enzymes into the bloodstream or removing bone marrow cells, than altering them and reimplanting them intravenously, with the hope that they will reproduce.

Nobody should underestimate the complexity of the process. The alteration of cells and bone marrow in laboratories is a demonstrated achievement. Their reintroduction in the human body is more difficult. There vectors must guide them to the right place in human cells. Nobody now knows how to excise the harmful genes, but perhaps the modified ones can overrule them. Then the cells must accept them, integrate them, and turn them on. All this is not easy, and a lot can go wrong.

The National Institutes of Health gave permission for the first cautious experiments in gene therapy in 1990, under the direction of Dr. W. French Anderson, who had long been planning for such an event. Within five years more than sixty efforts at gene therapy were underway. There have been no reported disasters and some encouraging outcomes. The results are sometimes described as "obscure," sometimes as "promising."

By the time this book comes off the press, there may be different news. And the media may report new developments at any time.

5. Germline therapy

The genetic therapies I have seen describing affect the somatic cells of the body. If successful, they may bring welcome healing to individuals. But those individuals may still pass on their ailments to their children. The reason is that the germ cells, the sperm and ova that are a minute part of the organism, have a life and history of their own, in some ways unaffected by the far more numerous somatic cells.

Inevitably, people are asking, with breathtaking eagerness: Might it become possible to treat the germ cells and remove a dangerous genetic trait from the children and future descendants of the parent originally afflicted? Two possible methods are proposed.

The first is to genetically modify sperm or eggs before fertilization, then arrange *in vitro* fertilization and implant the fertilized egg in a woman's body. Test-tube babies are already a present reality. The proposal for genetic altering of sperm and/or eggs would add a startling operation to the process.

The second, more likely method is to genetically modify the fertilized egg very early, before cell differentiation. As the fertilized egg divides and redivides, differentiation begins—at the eight-cell stage, it is sometimes said. Up until that time, any

Roger Lincoln Shinn

alteration affects all the cells equally, including those that will eventually become germ cells.

This is the possibility (maybe, impossibility) that seizes the imagination of some scientists and the wider public. It would be a stupendous act. Controversies, both scientific and ethical, rage around it. They focus on so many issues that I shall give them a chapter of their own (Chapter 6).

Keeping Up to Date

Every book about present-day genetics, including this one, is ready for revision by the time it reaches readers, as I have already suggested. Science is moving faster than publishers can produce books. For specialists, there are excellent technical journals. But even the specialist who wants the latest developments must sometimes rely on the same mass media that inform the rest of us. The press, radio, and TV are often helpful, often misleading.

Science reporting has improved immensely over the past two decades. But the reports require carefully reading. They often simplify complex phenomena (as I have simplified in this book). That need not be a fatal flaw; all communication requires selection and simplification. So the question is: Does the simplification give insight into the reality or distort it? The problem is at its worst in headlines and spot announcements. These often mislead in two ways: (1) They are, of necessity, too short to tell a complicated story. (2) They are designed to grab attention.

Two examples have been analyzed by Suzanne Holland and Donna McKenzie (Holland and McKenzie 1994). The first concerns genetics and homosexuality. The issue arouses furious emotions, with some people hoping for evidence that genes determine sexual orientation and others hoping for the opposite. Some gay advocates, believing that homosexuality is not an aberration but one form of normal diversity, resent the widespread attention to the issue. In this atmosphere, Dean Hamer and four colleagues published results of their research under the heading, "A Linkage between DNA Markers on the X Chromosome and Male Sexual Orientation" (Hamer et al. 1993). In cautious phrasing they reported that, in a small sample, they found that men with a specific genetic structure were more likely to be homosexual than other men. And they concluded: "at least one subtype of male sexual orientation is genetically influenced." With scientific restraint they avoided the word cause, but repeatedly used the verb or noun influence. They never referred to "the gay gene," although Dean Hamer less cautiously entitled a subsequent book, *The Search for the Gay Gene and the Biology of Behavior* (Hamer and Copeland 1994).

My aim here is not to evaluate the quality of the research, which has been both praised and criticized. My interest is in the public reporting. The article immediately made a splash in the press. *Time* headlined: "Born Gay? Studies of family trees and DNA make the case that male sexuality is in the genes" (July 26, 1993, 36–38). *Newsweek* ran the heading: "Does DNA make some men gay? Science: the biology of destiny" (July 26, 1993, 59). The question marks in both headlines were a small saving grace, but the subheads were pronouncements, not queries; and *Time* omitted the question mark on its front cover. The articles themselves were more carefully written. Those who read them will get a reasonably sound report on the actual research. But those who read only the headlines and those who skim the articles under the influence of the headlines—whether they be hostile or friendly to homosexual males—will find their own attitudes confirmed. I take up other aspects of this issue in Chapter 5.

The second example, more briefly, has to do with women's breast cancer. The work of Mary-Claire King, followed by further work of other scientists, succeeded in identifying a gene for susceptibility to breast cancer. The discovery deserved headlines and got them. But again the details are important. The experimenters estimated that only 5–10% of cases of breast cancer in the United States are hereditary, and only half of these are related to the newly-discovered gene. Among those who have it, the great majority—but not all—will get breast cancer. The true importance of the discovery, both for science and for women at risk, is immense. But it does not justify exaggerated reports.

So my advice is: Follow the news reports in the media. They are indispensable. But read, look, and listen carefully. Don't stop with the headlines or the spot announcements.

This chapter, I hope, is an orientation to what follows. With many an omission and many a shortcut, it surveys a wide terrain. Even this hop-skip-and-jump method gives a glimpse of the dramatic new human powers that require ethical decisions.

Chapter Three

NATURE, NURTURE, AND FREEDOM

Human freedom is both an aspiration and an enigma. We often affirm it, struggle for it, acclaim it. Even so, we sometimes realize that we are driven by forces of nature and culture that we did not choose. People often reject freedom, perhaps because they fear it, or because they want to avoid responsibilities for their actions, or because some scientists and philosophers tell them it is an illusion. But the very denial, we have to believe, is an affirmation of a free mind.

The eminent theologian H. Richard Niebuhr once wrote:

> Here we are, "thrown into existence," fated to be. How far our freedom may modify the qualities of our existence we do not know; we know we have some freedom, but our freedom is not our beginning. One thing we understand perfectly and that is that we did not exist ourselves into being and that however we may change the qualities of our bodies or of our minds, this self which lives in this body and this mind did not choose itself (H.R. Niebuhr 1989, 65).

Roger Lincoln Shinn

Those words, although published posthumously in 1989, were written in the 1940s or 1950s, probably before the discovery of the double helix and the outburst of public attention to genetics. But they take added force with our present awareness that many qualities of the "free" self are influenced by the jumbled alphabet of **A,G,C,T**. We—all of us—are determined by nature in ways that we do not entirely understand. Yet the ability of a self to make genuine decisions is a prerequisite to all ethics. When a car crashes into a pedestrian, we do not say that the car was unethical. We attribute the crime or accident to a driver or car manufacturer or mechanic. Cars do not make decisions; people do.

Nature and Nurture

A starting point for discussion is the old, old discussion of the roles of heredity and environment, of nature and nurture in shaping human life. Heredity, beyond any doubt, has something to do with every person's constitution, physical and mental. Environment, equally beyond doubt, is also a force. The relative contributions of each are not clear.

One of the sayings attributed to Yogi Berra, baseball star and popular sage, runs: "Baseball is 90 percent mental. The other half is physical." In a comparable way, some biological scientists argue that 90 percent of personal qualities are the inherited nature, with the "other half" environmental nurture. Many social scientists and educators prefer to reverse the figures, accenting nurture.

It may be that the important issue is not the quantitative weight that we give to heredity and environment, but the peculiar way in which the two interact in human life. Sometimes a person rises to the demands of an occasion, matching unique inborn potentialities with the unique need of the hour, in ways that could never have been predicted. Winston Churchill, in a more placid time, might have been a near nonentity in world history, a bumbling Colonel Blimp, enjoying his political maneuvers, his drinks, and his amateurish military abilities. Martin Luther King, Jr. might have remained a conventional upwardly-mobile black pastor with a fondness for orotund phrases and a weakness for plagiarism. In each case, a historical crisis struck sparks that ignited inner potentialities and changed history.

That opinion does not set aside the issues of nature and nurture. Each side in the debates can produce some scientific evidence to support its case. But the argument is scientific only in part. People choose their scientific evidence to fit their preferences. Authoritarian ideologies tend to emphasize nature, because they are friendly to unchangeable hierarchies of merit and status. The traditional justifications of caste and slavery rested on beliefs in a preordained nature of people and social groups. Doctrines of the divine right of kings or of oligarchies did as well. Liberal democratic ideologies tend to emphasize nurture, with the hope that education and culture can go far to overcome existing inequalities. My own convictions at this point are democratic, but that does not relieve me of the responsibility to examine the evidence for the very real determining power of nature.

The new genetic knowledge, although it does not put all the weight on nature, tends to shift the balance in that direction. It shows that many characteristics, whether physical or mental or psychic, are genetically influenced. DNA influences one's health, one's mental traits, and, yes, one's character.

At their most radical, some geneticists see a person as only a set of electrochemical processes. All human activities—all freedom, all personality, all art, all morality, even all science—are reduced to mechanistic, chemical processes. James Watson sometimes edges close to saying that:

Roger Lincoln Shinn

> I have spent my career trying to get a chemical explanation for life, the explanation of why we are human beings and not monkeys. The reason, of course, is our DNA. If you can study life from the level of DNA, you have a real explanation for its processes (Watson 1992, 164).

Even more emphatically, Thomas F. Lee writes: "We follow the orders of DNA. We have no choice. We are prisoners of our genes" (Lee 1991, 6).

But neither of these scientists wants to remain a prisoner of his genes. Watson goes on to make a strenuous appeal for attention to ethics, including laws "to prevent genetic discrimination and to protect rights that should not be signed away too easily" (Watson 1992, 172). Here he does not support his argument by offering a chemical explanation for it; he is concerned with human worth. Even more strikingly, Lee decides that "not a scientist worth his or her salt would ever suggest that people, him- or herself included, are just 'bags of genes.'" He affirms that "organisms are more than just DNA and the proteins for which it codes." And he sees "increasing control over our genetic fate." The final paragraphs of his book are an appeal for "loving, creative lives" rather than destructiveness (Lee 1991, 226, 246, 207, 300–301). When scientists write books, they do not say that their DNA is doing the writing.

Turning now to the advocates of nurture, we find a parallel problem. The pioneer American behaviorist, John B. Watson, all but denied the role of heredity in his insistence that a competent psychologist can do almost anything with human nature. His famous boast was: "Give me a dozen healthy infants, well-formed, and my own specified world to bring them up in and I'll guarantee to take any one at random and train him to become any type of specialist I might select—doctor, lawyer, artist, merchant-chief and, yes, even beggar-man and thief, regardless of his talents, penchants, tendencies, abilities, vocations, and race of his ancestors" (Watson 1930, 104).

Watson's most prominent American successor, B.F. Skinner, argued that the human genetic endowment was primarily an ability to respond to environmental stimuli. The environment, physical and social, gives positive or negative reinforcement to various behaviors. His belief about freedom is expressed in the title of his best-known book, *Beyond Freedom and Dignity*, 1971. The "feeling of freedom" is deceptive, he said, because "freedom is a matter of contingencies of reinforcement." A scientific account of behavior can ignore "personalities, states of mind, feelings, traits of character, plans, purposes, intentions" (Skinner 1971, 37, 15).

Skinner thus ran into the same problem that we have seen in any thoroughgoing genetic determinism. He did not apply his theory to himself. He said: "We sample and change verbal behavior, not opinions." But he wrote a book, clearly trying to persuade people to change their opinions. His final sentence was: "We have not yet seen what man can make of man"—not, "We have not yet seen what environmental reinforcements, themselves determined by other environmental reinforcements, will make of man" (Skinner 1971, 95, 215).

Three years later, Skinner reaffirmed his cheerfulness (Skinner 1974). But in an interview seven years after that, he had changed to despair. "I'm very pessimistic. We're not going to solve our problems, really." (Greenberg 1972, C1). We can put aside the questions of his changing mood, since Skinner believed that feelings are at best by-products of behavior. The point is that, in the debates over nature and nurture, the claimed denial of freedom can be as complete among environmental determinists as among genetic determinists.

There is no necessary affirmation of freedom in the comingling of two kinds of determinism. The most that can be said is that each form of determinism undermines the absolute grip of the other form of determinism. But an affirmation of freedom requires another step.

DETERMINISM, CHANCE, AND FREEDOM

For centuries people in many cultures have struggled with issues of determinism and freedom. Hindu beliefs about karma and caste, the biblical faith in divine providence, the Platonic hierarchy of human types, Stoic philosophies of nature, tribal and racial doctrines of human worth—all have recognized major determinants on individuals and communities, but have still insisted on some form of human responsibility for character and action.

Modern science accentuated the problem. From the seventeenth century until recently, scientists usually interpreted the cosmos as a vast deterministic system. René Descartes, for example, saw nature as a mechanism, obeying inexorable laws. Within it, the human body was also a mechanism. Struggling to sustain some human freedom and dignity, he developed his theory of dualism, fateful for many who followed him. Somehow, he believed, the human "mind" or "soul" was immune to the laws of physical nature. It could think, will, affirm and deny, and imagine (Descartes 1641, No. 1). The difficulties in that dualism were insuperable, and it could not endure. The biblical tradition had always insisted on the historicity of human beings—their rootedness in time and place, in communities with memories and expectations. The modern behavioral sciences gave evidence that there is no human self that stands outside the processes of biology, unaffected by them.

Even so, it is easier to believe in freedom now than in most of modern scientific history. Werner Heisenberg's theory of indeterminacy marks a scientific revolution as epochal as Einstein's theory of relativity. It affirms that something unpredictable, something akin to spontaneity, goes on within the atom. That does not guarantee freedom. A combination of necessity and chance does not, of itself, add up to purposive action—a point argued strenuously by Jacques Monod, the French scientist, in his famous book *Chance and Necessity*. But the theory of indeterminacy, without assuring freedom, breaks the iron vise of necessity. Interestingly, Monod himself, in his final pages, pleaded for an "ethical choice" and called upon "the highest human qualities, courage, altruism, generosity, creative ambition," recognizing both "their sociobiological origin" and "their transcendent value" (Monod 1970, 176, 178–79). I find it impossible to believe that Monod himself, the musical conductor and organizer of a Bach society and a daring activist in the French resistance to Nazism, was merely the outcome of the interplay of chance and necessity—like, say, an extremely complicated game of chemical craps.

The explosion of genetic knowledge, already far surpassing what Monod knew or expected, has a paradoxical relation to freedom. On the one hand, it is scientific knowledge of constraints. It informs us that our lives are determined by intricate physical-chemical processes involving complex proteins, base pairs, that alphabet of **A,G,C,T** that spells out all the genetic possibilities of all human beings. The molecules in those processes are not trying to do us any favors. They know nothing about us, yet they determine much of our nature and action. Drastic illnesses, produced by aberrant genes (aberrant from our point of view, not theirs, since they have no point of view) destroy many of our powers of choice and action. Even apart from diseases, the mindless processes of genetic chemistry influence our individuality in ways that we scarcely suspect. Modern genetics makes it more obvious than ever that there is no human self that stands outside the processes of biology, unaffected by them.

Roger Lincoln Shinn

On the other hand, genetics suggests the possibility of greatly enhanced freedom and power. The advocates of an aggressive genetics hope to surmount some of the constraints and illnesses that have plagued human life from the beginning. They never write as though they themselves were mechanically obeying the commands of the impersonal acid in their genomes.

How, then, shall we understand the enigma of freedom? Perhaps, as I am inclined to think, it will always remain a mystery, calling as much for reverence as for explanation. But we can think about it and, in particular, make some beginnings of appreciating its relation to chance and necessity.

By freedom I mean the ability of a person to envision alternative possibilities, make choices, and act to realize those choices. It does not mean that human behavior is uncaused. Yet it intrudes foresight and purpose, as causative forces, into the narrow causal systems that human minds have discerned in nature. It means that a free person is able, in some incalculable way, to act on the basis of purposes that modify the binding reign of mechanical causality or blind chance. Foresight and expectation are causative, as truly as unknowing processes of matter in motion.

Obviously freedom is not total. We live with physical constraints: we are not free to flap our arms and fly with the birds; we are not free to prolong physical life forever. We live also with social constraints: in human communities, personal freedom must be related to the freedom of others, who may support our freedom or block it. Extravagant claims for freedom—perhaps those in the early writings of Jean-Paul Sartre—are braggadocio. To repeat H. Richard Niebuhr's words, "We did not elect ourselves into being."

Yet we do envision unrealized possibilities, then we make choices and direct actions toward goals. We can even guide our conduct by "transcendent" ideals, to use a language that has a long history in philosophy and religion but that, at this moment, I borrow from Monod. We do not have to confine life to the very real, but not omnipotent, struggle for the survival of the fittest. We can subordinate our own interests to a larger good.

There is a kind of reductionism that wants to explain all altruism as dictated by "selfish" genes. The gene acts—unknowingly, of course—to perpetuate itself, not the person in whom it resides. If parents sacrifice themselves for their children, the reason is that the survival of the parental genes—those that have been passed on to the children—is promoted far better than if the parents abandoned their children for the sake of parental survival. Richard Dawkins seemed to say that on some pages of his provocative book *The Selfish Gene*. But those who read his book to the end—not all of those who argued about it—discovered his belief that persons, unlike genes, have "the capacity for conscious foresight." To that, he added: "It is possible that yet another unique quality of man is a capacity for genuine, disinterested, true altruism . . . We, alone on earth, can rebel against the tyranny of the selfish replicators." A second edition of the book, thirteen years after the first, retains and heightens that theme in its new endnotes and two new chapters (Dawkins 1989, 200–201).

Two examples help to suggest how freedom can influence the operations of nature. I intend neither of these to be a proof; I doubt that freedom can ever be proved or disproved. But the examples show ways in which freedom, related to systems of determinism and chance, still makes sense.

The first is from the most ordinary sort of experience. When I stand, relaxed, my arms fall to my sides, in accord with the laws of gravity. I can, and frequently do, raise an arm. I may do so for many purposes: to pick a cherry, to serve a tennis ball, to comb my hair, to swat a mosquito, to vote in a meeting, to point to a bird, to put my luggage in the overhead compartment, to wave at a friend, and so on and so on. In no case do I violate the laws of gravity. But in no case do the laws of gravity, of themselves, determine or predict my action. The explanation from the bottom up—the impersonal processes of chance and necessity—is not annulled, but it is inadequate without an explanation from the top down—an explanation that requires recognition of purpose. The radical behaviorist, of course, will reply that purpose is an illusion and that a physical explanation can be given for all these acts. But behaviorists themselves act purposefully when they design their experiments and argue their case.

The second example is more technical. It comes from genetics. All the cells of the human body, except for the few germ cells, contain the entire genome of an individual. The cells in the toes contain the genes for seeing; the cells in the intestines, the genes for the brain; the cells in the backbone, the genes for the heart. But the genes for seeing are not "expressed"—or "turned on"—in the toes, as the genes that grow toe nails are not expressed in the eyes. (In 1995 scientists reported an experiment in which they turned on the genes for eyes in various parts of fruit flies, producing eyes on wings and legs. Unassisted fruit flies turn on the genes for eyes at more appropriate places. The scientists did not propose to try the experiments on human beings.)

The human body would be a horrible mess if all the genes expressed themselves all the time. The relation of expression to bodily location means that there is no simple determinism operating from gene to function. The larger organism, within which the gene exists, influences the activity of the gene. I give orders to my genes as truly as they give orders to me. It appears that there are genes that turn on or turn off the operations of other genes. In that sense the bottom-up explanation has its validity. But that turn-on or turn-off operates in the interests of the encompassing organism. That is a top-down explanation. Again, I do not claim that this "proves" freedom. It shows that the activities of the genes are not simply self-enclosed. The chemicals act one way in a test tube. They act differently in a person and act many different ways in the same person.

The bacteriologist René Dubos, after looking at the evidence for determinism, replied: "In practice, however, all human beings, including the most deterministic philosophers and experimenters, believe that they have some measure of freedom in their decisions or at least in their choices; freedom posits free will" (Dubos 1972, 75). Similarly, the eminent geneticist Theodosius Dobzhansky wrote: "The ability of man to choose freely between ideas and acts is one of the fundamental characteristics of human evolution. . . . Ethics emanate from freedom and are unthinkable without freedom" (Dobzhansky 1956, 134).

The Oddity of IQ

Controversies about nature, nurture, and freedom occasionally explode into noisy public quarrels, usually connected with differences that are already sensitive. Some of the inflammatory issues are race, gender, economic status, and intelligence. Here I take a single example for examination: the furious arguments sparked by *The Bell Curve*, a hefty book by psychologist Richard J. Herrnstein and political scientist Charles Murray, 1994 (hereafter H/M). Disputes raged in the press, TV talk shows, schools, clubs and bars, and private meetings in corporate headquarters—usually by people who had not read the 800-plus pages of the book. One reason that the book was incendiary was that both authors already had reputations. Herrnstein, for more than two decades, had been known for his assertions about an inherited black inferiority (Herrnstein 1971). Murray had publicly opposed government welfare programs, food stamps, and similar efforts to help the poor (Murray 1984). Critics charged that many people welcomed the new book because it gave respectability (backed by elaborate charts and statistics) to their own opinions that they had been ashamed to proclaim openly.

As a starting point, it helps to demystify the bell curve. It did not descend from heaven. It is not a law of nature, to be applied to unknown situations, as we assume that the laws of the lever work for levers we have never seen. It is a generalization about some evidence some of the time. A look at the data shows that the distribution of some measurable characteristic in a given population (people, animals, plants), when put on a graph, frequently but not always takes the shape of a bell. The perfectly symmetrical bell, pictured on the dust jacket and some pages of H/M, is an idealized symbolic norm. (Remember the observations in Chapter 2 about statistical norms and reality in the flipping of pennies: the pennies rarely "obey" the statistical laws exactly.) In real life, bell curves are derived from the evidence, not imposed on it. In general, evidence from large populations tends to approximate the idealized curve, whereas small populations are more idiosyncratic.

In a bell figure, the horizontal dimension represents any measurable phenomenon. Take human height as an example. The left end of figure 1A represents the shortest people; the right end, the tallest. The vertical dimension represents the number of people of each height. In any group there is only one shortest and one tallest; most people are closer to an average.

Figure 1A is an idealized, theoretical bell curve. Actual curves, based on evidence, have their lumps and variations, as in figure 1B. It, too, is a convenient symbol; it is not based on any specific evidence, but simply illustrates the irregularities on any graph that represents the real world. From now on, all my drawings, even though theoretical, will show some lumpiness, simply as a reminder that this wonderful and confusing world is not easily graphed. Many of H/M's diagrams are equally lumpy.

Let us continue with the example of height, because it is

Nature, Nurture, and Freedom

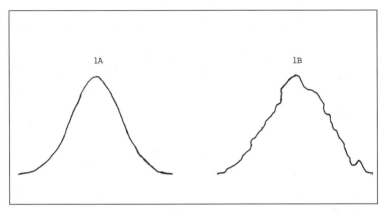

Figure 1

relatively noncontroversial—that is, relatively free from the anger, prejudice, and resentment that confuse the arguments about IQ. As a rough-and-ready case, somebody might chart the bell curve representing the height of the students in a given high school. Without filling in exact figures, which will be different for every school, we can expect something like the bell curve in figure 1B with the new shortest and the few tallest, and with most students in between.

Within the school are many identifiable groups. We might separate out freshmen, sophomores, juniors, and seniors. The tallest freshman might look down on the tallest senior, but that is not likely. Usually the seniors will average taller than the freshmen. We could expect four overlapping bells, for the four classes, as in figure 2A. We could measure boys and girls separately and get two overlapping bell curves. It is possible, although not likely, that the tallest student in the school would be a girl; in general, we expect the boys to be, on average, a little taller. We might separate out the basketball teams. The girls' team might come out a little taller (in figure 2A a little farther to the right) then the general population of

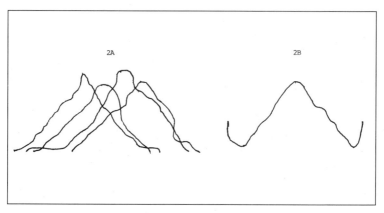

Figure 2

the school, but a little shorter (to the left) than the boys' team. But the tallest girl player might, contrary to averages, be taller than the tallest boy. There's no way of knowing without a look at the evidence.

We could similarly graph—somebody probably has graphed—any characteristic that can be measured, not simply among high school students but among any group that can be located and studied: weight, sleeping hours, time spent watching TV, money spent on this or that, drinking habits, and so on. We won't always get a bell curve. Drinking habits are an obvious example. In an identifiable group, the curve might be something like figure 2B. At the left are the teetotalers, who may outnumber the chablis-twice-a-year group. Then may come a sort of bell curve of moderate drinkers and heavy drinkers, trailing off to occasional binge drinkers, with maybe a small rise at the end of compulsive drinkers. All this, I repeat, might be; the actual curve depends on the group investigated. What is obvious is that the bell curve does not cover all cases.

What H/M did was to investigate IQ, relate it to the bell

curve, make an argument about its relation to nature and nurture (heredity and environment), and correlate it with race and class in the United States. Most of that was an old, old story. But H/M did it with a publicity and a book whose bulkiness—which might or might not mean thoroughness—seemed intimidating. To examine their argument, in relation to genetics and ethics, requires a sorting out of several issues.

1. What is IQ?

It's easy to talk about IQ (the standard abbreviation for intelligence quotient), but it is hard to say with any precision what it is. It is clearly not a measure of human worth. It is not character, not artistic imagination, not skill in human relations, not leadership. It is not motivation, diligence, or persistence. We would not try to judge the worth of such history-making characters as Franklin D. Roosevelt, Winston Churchill, Joseph Stalin, and Adolf Hitler by ferreting out their scores on IQ tests. We haven't the faintest knowledge of what Mendel's IQ was. We know that he failed his university examinations, and we don't know whether he'd have done more or less for genetic knowledge if he had passed them. Albert Einstein could not speak fluently at the age of nine, was expelled from school at sixteen, and later failed the entrance examinations for the Federal Institute of Technology in Zurich. After he eventually got his doctorate and achieved some scientific breakthroughs, his work was attacked by two Nobel laureates as "Jewish physics."

Nobody proposes to do away with elections and choose winning candidates by testing their IQs. Corporations do not choose their executives and colleges do not choose their presidents that way. So why do we care about IQ? One answer comes from economist Arthur Okun: "Stress on IQ is a form of narcissism peculiar to intellectuals" (Okun 1975, 85).

Roger Lincoln Shinn

Yet the idea of IQ is not *totally* meaningless. IQ is a measurement, of sorts, of a certain kind of mental ability. If we ask what kind, the only sure answer is: the kind that is measured by IQ tests. The Graduate Record Examination, used by universities in the United States, has three sections: verbal, mathematical, and analytic; students often score radically differently on the three. It does not purport to be an IQ test, but it shows something about the variety of mental aptitudes and skills. Even so, the common opinion that some people are "smarter" than others is not entirely mistaken. Psychologists sometimes refer to q, a general intelligence that is more or less an equivalent of IQ. And IQ has some relation, far from exact, to ability to cope with problems in our society.

I must emphasize *our* society. There is no evidence, genetic or historical, that we today are any smarter than our ancestors long ago. The unknown genius, or geniuses, who invented the wheel may have been as intellectually keen as a Newton or Einstein. But he or she or they lacked the ability to balance a checkbook, make a phone call, probably read and write. All these are cultural achievements. To ask what those predecessors of us would do with a No. 2 pencil or a computer on one of our IQ tests is silly. Testing experts try to devise tests that are culture-independent, but they really don't know how.

These considerations do not require us to throw away the notion of IQ as a useless fraud. Indeed, IQ tests have sometimes had a liberating effect for *some* people. In societies where inherited rank confers power and status, educators have hailed IQ tests that discern youths in the lower classes who have greater potentials than the drones in a decadent aristocracy (Wooldridge 1995). But the same tests have been used to classify children and deny opportunity to those less adept at taking such tests. We had better approach the concept of IQ without dogmatism but with suspicion.

2. Nature, Nurture, and IQ

Almost nobody would deny that heredity has something to do with IQ. How much? Nobody knows. H/M, plucking a figure in the midrange of current estimates, guess 60 percent.

A few things are obvious. I am sure that, with the best physical nurture in the world, I could not have been a Babe Ruth or a Michael Jordan. I don't have the genes for their mighty deeds. With the best artistic nurture, I could not have been a Rembrandt or a Picasso. With the best intellectual education from infancy, I could not have been a Newton or an Einstein. But Newton did not know how to maneuver a car through the interchange on a superhighway, or select foods to lower his cholesterol, or use the word processor on which I write this book. If he were to appear in our society, after overcoming his culture shock, he could learn those techniques, probably faster than I. But, then, he would be a different Isaac Newton. We are all expressions of our culture as well as of our genes. In the interaction of the two, we become the individuals whom we are.

3. Is there any connection between IQ and race?

To begin, scientists are increasingly skeptical that anybody knows what race is. The most thorough recent studies by population geneticists have produced a genetic atlas, which shows greater genetic differences within races (as that term is generally used) than between races. Luca Cavalli-Sforza and his colleagues conclude: "There is no scientific basis to the belief of genetically determined 'superiority' of one population over another" (Cavalli-Sforza, Menozzi, and Piazza 1994, 19).

Here I had better say openly that I *want* to believe that and find the evidence for it convincing. For that reason, I have a special obligation to investigate the arguments of those who disagree. So suppose we work with the conventional notion of race in our society. Then there are, at this time, some measurable differences between the average or composite skills of different groups in taking IQ tests, as there are in economic standing, educational achievement, political success, and some other characteristics. If the groups are sorted out, they fit some overlapping bell curves, on the general pattern of figure 2A. (That, to repeat a point, is a generalized all-purpose diagram, not a specific representation of IQ scores.)

H/M here report several findings in the United States today. White Americans, on the average, score higher than African-Americans. East Asians (Japanese, Chinese, and Koreans) score higher than white Americans. Ashkenazi Jews score highest of all.

Does heredity have anything to do with this? The only sure answer is: Nobody can be quite certain. Genetic heritage, history, and culture have melded so intimately and intricately in American society that nobody knows how to disentangle them. The most obvious observation is that, in a nation inaugurated with a Declaration of Independence—"all men are created equal"—the ethical meaning of intricate statistics about IQ is nil. Glenn Loury, University Professor at Boston University and a black social scientist, responds to H/M: "I would have thought, and have always supposed, that the inherent equality of human beings was an ethical axiom, not a psychologically contingent fact" (Loury 1994, 13).

After hammering down that point, I see four seasons to suspect any generalizations about the heredity of definable groups. The first is that, in the length and breadth of human history, brilliant innovative insights have come in remarkable diverse ethnic settings. Daniel Boorstin rightly calls attention to "the fluidity and many-sidedness of past experience" (Boorstin 1975, i). The early achievements in astronomy and the development of a calen-

Nature, Nurture, and Freedom

dar took place independently in China, India, Babylonia, and Egypt. In mathematics Egypt and Mesopotamia apparently led the world in the second and third millennia B.C.; later the Greeks did superb work but were held back by the lack of the number zero, which was discovered (or invented) by Hindus and brought to Europe by Arabs. Quite independently, the Maya of Central America also discovered zero. The Romans were good at construction, but their science was retarded by a clumsy system of numbers; Arabic numbers, imported into Europe, liberated mathematics. Writing developed independently in Egypt, Mesopotamia, China, and among the Maya. The Chinese invented block printing in the seventh century. India led the world in cataract surgery at about 500 B.C. Africans learned to work iron before northern Europeans. Polynesians apparently developed amazing navigational skills, quite on their own, in the early centuries A.D. If Europe led the great age of modern science, nobody knows where innovators will appear in the third millennium.

A second reason emerges out of a peculiarly American experience. In 1943 Ruth Benedict and Gene Weltfish published a booklet, *The Races of Mankind*. Inquiring into the relation between heredity and culture, they investigated intelligence tests given to the American Expeditionary Force in World War I. This may have been the biggest mass testing, using identical tests, on record. In those tests Negroes (the preferred term when Benedict and Weltfish wrote) averaged lower scores than whites. But northerners averaged higher than southerners. Benedict and Weltfish came up with the data, devastating to some people, that northern Negroes, on the average, scored higher than southern whites. "Everybody knows," they said, that southerners and northerners are "inborn equals." They attributed the difference to diet, housing, income, and expenditure on schools. They concluded: "The scientist realizes that every time he measures intelligence in any man, black or white, his results show the intelligence that man was born with *plus* what happened to him since he was born" (Benedict 1950, 182).

The history of the pamphlet reveals something about our society. The YMCA, the USO, and the Army Morale Division began to use the pamphlet among soldiers in the World War II Army, in an effort to overcome prejudice. When Congress discovered that, the fat was in the fire. A special House Military Affairs Subcommittee found "All the techniques . . . of Communistic propaganda" in the pamphlet, and the army decided to leave it alone (Benedict 1950, 167).

A third reason has to do with recent discoveries about infant care and development. This is mentioned in passing by H/M, but not developed (H/M 1994, 389). The Carnegie Task Force on Meeting the Needs of Young Children has gathered and publicized important data. In the prenatal period the brain grows from a few cells (or neurons) to billions. In the months after birth the synapses (connections between neurons) increase from fifty-six trillion to 1,000 trillion. Inadequate nutrition in this period can lead to learning disabilities and mental retardation. "Studies of children raised in poor environments—both in this country and elsewhere—show that they have cognitive deficits of substantial magnitude by eighteen months of age and that full reversal of these deficits may not be possible." The structure of the brain, metaphorically called "the wiring," is in significant degree a continuing process in response to people and sensory experiences. Diet and family relations are critical in the whole development (Carnegie Task Force 1994, 7–8). To sum up, intelligence is not simply a given in the newly-fertilized egg and there determined for a lifetime; it develops. Deprivation in infancy and childhood rob children of "innate" potentialities. In a society where "African American babies are twice as likely to die within the first year of life as white babies," it is clear that many of the survivors suffer continuing physical and mental disadvantages (Carnegie Task Force 1994, 4).

A fourth consideration concerns a contradiction in H/M, which they realize but then neglect. It is their belief that American society has been deteriorating in IQ, through most or all of this century, at the rate of about 1% per generation (H/M 1994, 341–48). They attribute the decline both to demographic patterns (higher birthrates among people of low intelligence) and immigration. Yet they report "the Flynn effect"—the data that IQ scores on standardized tests have been rising, in this country and elsewhere, since World War II (H/M 1994, 307–309). That throws in doubt their whole thesis.

4. What recommendations come from the data?

H/M want to fend off the caricatures that they invite. Occasionally they assure us that they believe All-the-Right-Things. They say, even if they sometimes forget, "the concept of intelligence has taken on a much higher place in the pantheon of human nature than it deserves" (H/M 1994, 20–21). They know that "differences among individuals are far greater than the difference between groups." They tell us: "It should be no surprise to see (as one does every day) blacks functioning at high levels in every intellectually challenging field." And later, "A person should not be judged as a member of a group but as an individual." They endorse ways of finding dignity and adequate income for all groups in a society where everybody is "a valued fellow citizen" (H/M, 270–71, 278, 550, 551–52). It is only fair to recognize such declarations. But it is also fair to point out that, in context, they become pious platitudes without much operational meaning. And they are almost swamped in the overall direction of the book.

The total effect is accentuated in an eleven-page article by Murray and Hernstein, "Race, Genes and I.Q.—An Apologia,"

Roger Lincoln Shinn

published in *The New Republic* after completion of the book (hereafter, M/H). Here they make the case that most people identify themselves with "clans," nourishing themselves with a "clannish self-esteem." They propose a hypothesis: "Clans tend to order the world, putting themselves on top, not because each clan has an inflated idea of its own virtues, but because each is using a weighting algorithm that genuinely works out that way." They endorse "wise ethnocentrism" (M/H 1994, 36, 37). Many black people, reading this, hear the message: Don't feel bad that your ethnic group lags behind in intelligence. Enjoy the camaraderie of your clan. Don't assume that the dominant white clans, just because they think they are superior, regard you as inferior. Be happy that you are better than they at jazz and basketball. And don't expect any government programs to remedy the inequalities that are inherent in society.

No, M/H don't quite say that. They mitigate slightly their clannishness by saying: "Americans often see themselves as members of several clans at the same time—and think of themselves as 100 percent American as well" (M/H 1994, 36). But the implicit logic of the larger book becomes explicit in the shorter summary, and it breaks through the occasional qualifications. It is hard to believe that M/H are so naive about the hostilities lurking throughout American society as to assume that their book could not have pernicious effects.

But before leaving M/H, I should express one agreement with them. They are concerned that our society is becoming increasingly stratified, with a "cognitive elite," who know how to operate in this high-tech world, and a group of outsiders, increasingly sunk in poverty and despair. The gap between haves and have-nots is growing. And the whole society suffers from unemployment, weakness of families, schools that fail to educate many children, and crime. Those are real issues. But this book is about ethics and genetics. One of its purposes is to deflate excessive reliance on genetic explanations for social woes.

5. Again, What about Freedom?

The discussion of IQ gives a poignant twist to the whole understanding of freedom. To whatever extent IQ is influenced genetically, it is conferred on us without our consent. It is no occasion for pride or shame. To whatever extent it is influenced by infant nutrition, family relations, and social institutions, it is a cultural product. That fact undermines the exaggerated individualism that runs deep in our culture.

Individuals cannot, by a sheer act of will, transform their IQs, just as they cannot assure themselves a healthy physique, a charismatic personality, or a long life. Yet IQ is not totally irreversible and, even more important, does not determine what any person becomes. This "narcissism peculiar to intellectuals," this idol in "the pantheon of human virtues" needs debunking. Purpose, motivation, and self-discipline influence what each of us does with his or her IQ. Our own acts and the acts of other people have some influence on IQ and far more influence on whatever we make of life. So we come again to the elusive but powerful reality of freedom, that gift and burden conferred on human beings without their consent.

The realm of freedom is the arena of ethical decision. We turn next to examine the processes of such decisions.

Chapter Four

SCIENCE, ETHICS, AND FAITH, IN PUBLIC POLICY

The launching of HGP required many decisions—by individuals, by institutions, by the United States Congress, by people and governments in many nations. These decisions were not prescribed by existing laws, habits, or traditions. They were new.

Here I intend to examine the nub of the issue of this book: the relationship of science and ethics in the making of such decisions. Although the decisions affect both personal careers and public life, I am centering the discussion on public policy.

I can start bluntly: I am making the claim, perhaps brash, that in all the diversity of issues and processes of public policy, there is a pattern that is universal and unavoidable. It is discernible in highly diverse forms of government, most obviously in democracies.

This is not to say that a government or body politic deliberately practices the pattern, as if trying to please academic scholars. The process is largely unconscious. We might compare it with blood circulation in a human organism. Blood does not circulate because people decide that it ought to do that. It did not begin to circulate when, late in history, William Harvey (1578–1657) discovered the process—a discovery partly anticipated by Michael Servetus (1511–53). Harvey described what had been going on for hundreds of thousands of years without his help. Still, there are advantages in knowing about blood circulation. Similarly, I see advantages in describing, as accurately as possible for a chaotic and imprecise process, the ways of societies in relating science and ethics to public policies.

In the midst of many varieties and idiosyncracies, certain constant elements and interactions enter into the working out of policies. Again, we can compare the circulation of blood. Individuals vary in their counts of red and white corpuscles, high or low blood pressure, cholesterol, triglycerides, high-density lipoproteins, plaque that blocks arteries, and a lot, lot more. But there is no human life without circulation of blood. And there is no public policy without a few identifiable factors.

I propose a model that is descriptive, not prescriptive. Again, as in the case of circulation of blood, description may lead to some prescriptions, and I do not shun them. However, the accent here is on method, not outcomes. Chapters 5 and 6 will move further toward conclusions.

How can I verify a claim for the validity of this model? I can explain it and invite anybody to verify or refute it from experience.

HGP is an appropriate case study for the model. Because it extends the frontiers of knowledge and power, it jolts us into awareness of what we are about. No scripture of any religion commands, "Thou shalt—or shalt not—cut and splice DNA." Nor does Plato or Aristotle or Spinoza or Kant or Bentham or Dewey. They never faced these decisions. Yet the new genetic discoveries do not themselves entail ethical affirmations or prohibitions. To make new decisions requires attention to perplexities of method that are less obvious in habitual decisions.

A Triangular Model

We can begin to represent the process by a triangle, with each vertex representing a center of attention. Each of the three is a complex gestalt, but conversation thrives on simple names. So the three can be called—very broadly speaking—values, science, and politics.

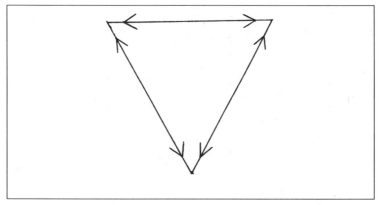

Figure 3

1. Values

A body politic approaches any decision with the interests, motives, purposes, commitments, ethical sensitivities, and religious beliefs of the society and of groups and individuals within it. For convenience I can use the pallid term *values* to represent that constellation of dynamic forces.

Values, minimally defined, are all things that are valued. As such, they embody the interests of persons and groups. All of us value necessities of life and personal advantage. But values may also transcend narrow interests and extend our horizons. They may lead us to seek a wider good than our own. They may evoke loyalties and purposes that override or revise our immediate interests. As Ian Barbour puts it, "subscription to a value also includes beliefs about benefits or moral obligations that can be used to justify or defend it or recommend it to others" (Barbour 1993, 26–27). I put values at the bottom point of the triangle, because purpose is the root of activity. But a point is not a flat base; the whole triangle is mobile and often tips in one direction or another.

What is the source of these purposes and values? The conventional answer is experience. But that answer avoids the crucial question: what experience and experiences? An experience of war leads some people to seek peace, others to try to build invincible military power. The experience of poverty may engender compassion for the needy or a drive to climb out of poverty on the backs of the poor. There is always something mysterious about commitments. Max Weber, although confessing that he personally was religiously "unmusical," insisted, in his famous essay on "Politics as a Vocation," that political commitment always rests on "some kind of faith" (Weber 1919a, 117).

But if faith, of some kind, is part of the answer, it is also part of the problem. In a pluralistic society, there are many faiths, many commitments. In part, our society meets this situation by letting individuals and groups do their own thing. But public policy must come out of some widely-shared concerns. A society can enjoy and appreciate many diversities, but government requires some measure of consensus—at a minimum, acquiescence. The Swedish Gunnar Myrdal, in his epochal study *An American Dilemma*, discovered a convergence of ethical insights in this multicultural society. He called it "The American Creed," and he found its roots in the Bible, the Enlightenment, and the British common-law tradition (Myrdal 1942, Chapter 1). That consensus in America is often strained, sometimes to the breaking point; and it does not always find response in an international consensus. Interests and commitments are diverse, sometimes bitterly or tragically so. They are not reducible to some neutral, non-ethical source.

In this book I speak without apology out of an ethic rooted in the biblical tradition—not the Buddhist or Platonic or Islamic or utilitarian, all of which deserve a hearing. But, entering into public conversation, I appeal to widely-shared values in the society at large.

To point to the diverse, nonreducible character of fundamental commitments does not mean that these are simply arbitrary and nondiscussable, as logical positivists used to say. Individuals and communities can examine their commitments. They can investigate the tensions within their own motivations and values, e.g., between freedom and equality, justice and mercy, truthfulness and compassion. They can listen to other voices and learn from other experiences. Facing new issues, as in the HGP, they can look for sources of wisdom anywhere. But eventually their judgments depend on some conception, never fully demonstrable, of human nature and destiny.

This became obvious in the early discussions of the new genetics. In 1980 the general secretaries of three religious organizations—the National Council of Churches of Christ, the Synagogue Council of North America, and the United States Catholic Council—joined in a letter to President Jimmy Carter. They asked for attention to "the entire spectrum of issues involved in genetic engineeering" (President's Commission 1982, 95–96). Carter assigned the subject to the already-existing President's Commission for the Study of Ethical Problems in Medicine and Biomedical and Behavioral Research. It produced a 115-page book, *Splicing Life: The Social and Ethical Issues of Genetic Engineering with Human Beings*.

Roger Lincoln Shinn

At the time of publication of that book, U.S. Representative (later Vice President) Al Gore conducted hearings of his House of Representatives subcommittee of the Committee on Science and Technology. He invited, among others, the familiar triad of Protestant, Catholic, and Jew. (I happened to be the person invited to testify on Protestant positions.) There was no expectation that the religious communities would determine government policy. But their testimony was sought. Why? One reason might be political expediency; any wise political leader wants to know the beliefs of significant constituencies. But, as the hearings made obvious, there was a genuine receptivity to ethical insights as the committee explored unmapped terrain (Committee on Science and Technology 1982).

Then with HGP came ELSI, the program on ethical, legal, and social implications. The Office of Energy Research publishes annual bibliographies, which by 1994 included cumulatively about 7,000 entries on ELSI (*Bibliography* 1994). The importance of one corner of our triangle was established.

2. Science

Values—the word I have chosen to represent that complex of interests, purposes, commitments, ethical sensitivities, religious beliefs, and motivations that shape decisions—do not of themselves produce policies. Every policy depends on some knowledge of the world we live in. I put that at the upper left point of the triangle. I use the term *science*, very broadly defined, to represent two elements of this worldly knowledge.

The first is verifiable information. Verifiable does not mean infallible. Scientific knowledge is always subject to revision in the light of further investigation. But verifiable knowledge is open to scrutiny and testing by a wide variety of people; it is many steps removed from arbitrary and capricious assertion.

Science, Ethics, and Faith, in Public Policy

Such knowledge is indispensable to the formation of policy. Think, for example, of the biblical prophet Amos. His voice resounds through the centuries as he declares the word of God: "Let justice roll down like waters, and righteousness like an ever-flowing stream" (Amos 5:24). That is the bottom point of our triangle. The name of Amos does not belong on any roster of scientists. But notice how much factual knowledge he reports about his society. He tells of war and exiling of people, of mistreatment of pregnant women, of selling the righteous for silver, of prostitution and drunkenness, of oppression of the poor. He assumes that this knowledge is verifiable by the observations of those who hear him. This knowledge is, in my present loose use of the term, scientific.

But science is not simply knowledge. An equally important element in it is the conceptual schemes by which investigators organize information and direct research toward the discovery and verification of new knowledge. We have our physical sciences, biological sciences, social sciences. In our era science has immense power. Its achievements require new decisions, most notably just now in medical sciences. Bioethics, so far as I know, was not even in the dictionaries before 1970. Soon it became the subject of a four-volume *Encyclopedia of Bioethics* (1978). It is a growth industry.

Genetics, for most of history a collection of folk observations, superstitions, and prejudices, is now a science. This is not to say that old errors have lost all their clout, but it is to say that verifiable information and conceptual schemes are bringing new understanding and new powers. In their train come new errors and superstitions that need purging in a continuous process, to which science contributes. HGP marks one momentous stage in that development. It is itself an expression of public policy and a catalyst for more public policies. Now, more than ever before, it is important to distinguish valid genetic knowledge from foolishness. Which of our illnesses derive from our genetic makeup, which from environment, which from accident, which from our own foolish behavior? And how do these various factors interact? Historical genetics was mostly phony. We are trying, in the midst of confusion and controversy, to get ours straight.

In our triangle, an individual or group might enter at either of the two points I have described: values and purposes, or scientific knowledge and power. But whatever the entry point, policy must soon move to the other point. Debates about public policy are partly debates about ethics, partly debates about scientific knowledge and expectations. It is important to recognize and sort out the differences.

Actually no person or community lives for long on either one of these two points of the triangle. We—all of us—live on a continuum, and societies make policy on this continuum. Responsible car drivers value mobility and safety; they need also accurate information about the flow of traffic, road conditions, braking on wet pavements, and a hundred other factual matters. Any debate about HGP involves multiple debates: some about scientific information, hypotheses, probabilities, and expectations; others about purposes and the human good.

The intriguing and puzzling aspect of the debate is the relation between the two ends of the continuum. I have sometimes said that the second most important issue in public ethics is to distinguish between the two. But that distinction is deceptive if it leads to the isolation of the two. So the most important issue is to see their relationship.

The distinction is utterly necessary. To confuse the two is to invite ethical and social chaos. No motivation, no passion will produce an effective policy without appropriating science (whether the direct, factual observations of an Amos or the highly technical findings and methods of geneticists). Motivation, although essential, is not an adequate answer to disease. We cannot, by simple moral fervor, end AIDS or the panoply of genetic illnesses. Any effort must pay attention to science.

This science has its own methods of experimentation and verification, and it rightly refuses to defer to external authority. Religious and political authorities have often erred, foolishly or disastrously, in trying to dictate scientific conclusions. That was the mistake of a papacy condemning Galileo, of fundamentalists refuting geological and evolutionary sciences by quotations from the Bible, of Soviet authorities prescribing the genetic theories of Lysenko. They did not respect the legitimate autonomy of science.

At the same time, science alone does not prescribe the human good. It can, within a social and ethical context, contribute to the good; but it cannot be the sole or final arbiter of the good. No pile of data, if it reaches to the moon, dictates an ethical decision or a public policy. A present example is the ability, unknown to earlier generations, to conduct tests of the human fetus. One of the easiest tests determines the sex of the fetus. That makes possible the abortion of a fetus of the undesired sex, a practice I have described in Chapter 2. If I believe that abhorrent (as I do), no scientific demonstration of the safety, reliability, and ease of the practice can, of itself, determine its morality.

Sometimes a moral consensus is so obvious that ethical debate is unnecessary. Some human ailments—e.g., cystic fibrosis, Huntington's disease, Tay-Sachs—are so devastating that nobody really wants them. Their disvalue can be taken for granted. Then, if a medical or genetic cure seems possible, the debate centers on its feasibility and risks. The factors of motivation and morality are not irrelevant; they are simply basic assumptions. An example is the changing ethic of smoking tobacco in this country. Scientific knowledge, almost of itself, instigated the change; but it did so because the value of physical health was, to most people, self-evident. However, in a world where little can be taken for granted—including the relative claims of personal freedom and public health, as in the case of tobacco—the ethical issues often emerge explicitly, although usually intertwined with the scientific.

My position has implications for the persistent argument about the possibility of a "value-free" science. This is an important debate, but it is often misplaced. Both sides in the argument make an important case and usually overstate it. Then both sides claim too easy a victory, because they refute a skewed representation of their opponents' case.

In my opinion, the most interesting debates on this issue, in their bearing on public policy, come from the sociologists of knowledge, because they direct their arguments toward public policy. Among social scientists, the most influential advocates of value-free science are probably Max Weber of a past generation and Peter Berger among contemporaries (Weber 1919b, 143–56; Berger 1963, 5–6). Their critics sometimes misunderstand them in three ways. (1) Neither is a positivist. Weber held that science requires as much "inspiration" as art, and he insisted that no science is presuppositionless (Weber 1919b, 136, 143–45). Berger is forthright in his rejection of positivism (Berger 1963, 168). (2) Both affirm that science assumes one primary value or constellation of values: the integrity of science itself, including the rigorous honesty that is essential to the scientific enterprise. (3) Both know that scientists, as persons in human communities, have values and that those values are important to their humanity.

Why, then, do they argue for a value-free science? Their point is that science requires an openness to evidence, whether or not it is pleasing to the scientist. There is a stubbornness about factual data; they do not always fit scientists' wishes or ethical preferences.

SCIENCE, ETHICS, AND FAITH, IN PUBLIC POLICY

Critics of value-free science sometimes argue that there are, in human experience, no uninterpreted facts. I agree, but that does not mean that there are no facts requiring interpretation. Weber insisted that the competent scientist (or teacher) give particular attention to the "inconvenient" facts (Weber 1919b, 147). Interestingly, Gunnar Myrdal, who advocates openly value-guided social science, nevertheless insists, "Facts kick" (Myrdal 1969, 40). Occasionally scientific error or fraud has burst into national attention when investigators skewed their evidence to support their favorite theories (Broad and Wade 1982). A recent conspicuous example was the claims for "cold fusion" of atomic nuclei.

What Weber and Berger know, but sometimes forget to say, is that science is a social enterprise, supported by industry, government, and foundations. The neglect is the more surprising in the case of Berger, since he and Thomas Luckmann wrote *The Social Construction of Reality*, a book describing astutely how human interests enter into the "construction of reality." Yet even this book insists, in its final paragraph, that social science, though "a humanistic discipline," is "value-free" (Berger and Luckmann 1967, 189). However, the social enterprise of science, which greatly influences the direction of research, is value-loaded. I have already noted (Chapter 2) how James Watson, describing the launching of HGP, made clear that the primary motivation of it, in his mind and the minds of the Congress that voted the funds, was a value-laden concern for human health (Watson 1992, 164–73).

Weber and Berger do not need to be told that human purposes and values influence the subjects and directions of scientific research. And the direction affects the outcome. Scientific investigators are more likely to find what they are looking for than what they are excluding from their attention. All concentration requires exclusion of distracting perceptions.

Granted, the unexpected and unwanted datum may appear, sometimes refuting a treasured hypothesis and redirecting attention to new possibilities. Also, serendipity operates in science: the unpredicted spot or fog on a photographic plate may lead to startling conclusions, more important than those the experiment was designed to discover. But the human genome can be mapped only by scientists seeking that objective in a society that values this knowledge enough to pour big resources into it.

However, for Weber and Berger the purposes and values of the investigators and the political and economic resources involved are not, in the strict sense, science. Their point is that the investigator who fudges the data to support a personal purpose or bias—even a "good" purpose or bias—is at that point not a sound scientist. Values may not intrude on the integrity of scientific discipline.

Because science is practiced by human beings, with common human fallibilities and prejudices, some human distortions frequently creep into purportedly scientific work. "Expert" scientific witnesses are usually available on both sides of public controversies—think of the costs and benefits of ballistic missiles or of a supercyclotron, or the merits of HGP—and, as it often turns out, the experts have a stake in the outcome. The public, suspicious of biased experts and especially of hired experts, seeks "disinterested" scientific testimony, which is often hard to find. To the extent that values skew findings, the "science" is unscientific.

A particularly interesting example of the relation of values to scientific neutrality is the work of Nancy Wexler, whom I mentioned in Chapter 2. As a young woman she learned, from her mother's suffering and death, that she was herself at 50% risk of the same severe ailment, Huntington's disease. That knowledge unquestionably shaped the daughter's professional life. Her illustrious career led her to arduous and compassionate research in Venezuela and in North American laboratories. It heightened her excitement as the Collaborative Research Group of six laboratories approached closer and closer, then succeeded, in the identification of the gene for the disease. It motivates her hope for future discovery of a remedy. There can be no doubt that this research is loaded with personal and ethical passion. That motivation contributes to its success. But at the point of evaluating the evidence discovered in the laboratories, Wexler, as an authentic scientist, cannot compromise scientific integrity to get the results she wants. She cannot claim an outcome that has not been accomplished. Her personal history, her identity as a woman, and her ambitions become irrelevant to her verification or refutation of the findings of her project. In that precise sense, and only in that sense, the research of this value-driven woman is value-free.

I have made the case for a kind of autonomy of both science and valuation as independent endeavors that cannot be confused without distorting both. But I have argued, equally strongly, that the two constantly interact in the formation of public policy. They represent two poles on a continuum. Policy is not settled at either pole. On the line connecting them, some policies approach one limit (either ethical or scientific) and some approach the other. A society cannot afford to neglect either pole.

3. Politics

The third point of the triangle represents the term *politics*. Although I come to it last, it has been there all the time. The scientist, austerely committed to exacting disciplines, may disdain its messy methods. The religious prophet may despise its wheeling and dealing. But neither can avoid it. Politics is the process for defining and implementing public policy. HGP is endorsed by government and funded by taxpayers' money.

Concerned citizens of a body politic are constantly asking two questions. (1) Which of our convictions do we want to legislate? Certainly, some; and certainly, not all. (2) Which is it possible to legislate? Again, some, but not all.

We can quickly dispose of the threadbare cliché, "You can't legislate morality." To that, I ask: What else is there to legislate? Apart from procedural details—American law tells drivers to keep right; British law, to keep left—legislation is a matter of rights and obligations; that is, of ethics. But some ethical beliefs are best left in the voluntary realm. And some are impossible to legislate.

If we think of the Ten Commandments, virtually all societies legislate some of them: "You shall not murder. . . . You shall not steal" (Exod. 20: 13, 15). Such legislation—despite glaring failures—is enforceable, more or less. What about the first commandment: "You shall have no other gods before me"? (Exod. 20: 3). Those most faithful to that commandment do not want to legislate it; they respect the freedom of other people to dissent. They also know that such a commandment is in principle unenforceable.

But if politics is a matter of ethics, it is also a matter of power. In the famous saying of Reinhold Niebuhr, "Politics will, to the end of history, be an area where conscience and power meet, where the ethical and coercive factors of life will interpenetrate and work out their tentative and uneasy compromises" (R. Niebuhr 1932, 4).

In some ethical talk, compromise is a shameful word. Of course, some compromises are shameful. But is there any public policy that is not a compromise? Are not democracies, above all forms of government, systems of compromise? Does not even family life involve compromise? How could any marriage survive without compromise?

HGP is possible because of an act of Congress and an appropriation of funds raised by taxation. Taxes, as all people know and most complain, are compulsory, not voluntary. What justifies taxing people for a project (HGP) that most of them do not understand? How do legislators decide to apportion revenues among national defense, law enforcement, and public health, along with many, many other projects? Within the health establishment, how should funds be divided among AIDS research, cancer research, HGP, and other claimants? These are political questions.

Then, when HGP gets underway, what is the role of government regulation? Government cannot prescribe its findings, without making nonsense of science. But government regulates the new powers that come out of the project. It authorizes or refuses to authorize certain patents. These are political decisions.

Human rationality enters into such decisions. The arguments about values and science that persuaded Congress, though debatable, were reasoned. Scientists persuaded legislators that probable benefits justified costs.

But those who follow the legislative process know that much more is involved. How do interest groups, power blocs, and lobbies get their influence? What trade-offs are necessary? What sways the crucial vote of a senator who happens, for reasons irrelevant to HGP, to chair an important committee? How do advocates maneuver through the system of checks and balances that often obstructs legislation? How do the prestige, the persuasiveness, the personal friendships of particular scientists influence specific legislators? What senators and representatives see money to be spent in their home states? Who has the ear of the President or a key staffer in the White House? In the case of HGP, who determines the outcome of the turf fight between the National Institutes of Health and the Department of Defense, both of which administer funds for research?

The whole process reminds us of Winston Churchill's quip that democracy is the worst of all forms of government, except for the alternatives. Its advantage is that it gives voice to many people with many opinions, and it has powers of self-correction. It is not a tidy process. It lacks the precision of scientific verification in a laboratory or of academic discussion among ethicists. And, for all its reliance on persuasion, it involves power. But politics is essential to big science in our time. And democratic politics offers more promise of justice than the alternatives.

Now, looking at HGP, we can see the indispensable relevance of all three vertices of the triangle. Society has made a major investment in a process of research that is producing knowledge and power never before known. The project comes out of the interaction of values, science, and politics. Each of these represents a different entry point to the great decisions that cry for attention. Various people and groups enter the triangle at different points. No one point can be reduced to or derived from any others. Yet no one point is independent of the others. All three are involved in the whole process and its outcomes.

Science opens up new possibilities, some for therapy and some, perhaps, for the modification of human nature. Neither ethics nor politics can determine what is possible when geneticists act upon DNA. But the possible is not always the desirable, and scientists cannot be the adjudicators of values for a whole society. The *values* of a society are not decisive in analyzing the composition of DNA or determining the possibility of a specific act of gene-splicing. Here society relies on scientific knowledge. But individuals and society will then evaluate that knowledge in terms of purposes and commitments that are not determined in laboratories. *Politics* is not a seminar in science or in ethics. But in setting public policy it takes account of scientific knowledge and potentialities, and it must (despite cynical doubts) be responsive in some measure to the values of the body politic. In particular, it must decide when to exercise its power of regulation and constraint, and when to hold back in order to permit freedom and initiatives in other organizations.

THE POWER OF IDEOLOGY

The model is still incomplete. I must put a ring around the triangle. The ring represents *ideology*. Policy always takes shape within an ideological context—or an ideological contest. The ring is not a perfect circle. It is shaped more like a cloud with indeterminate boundaries.

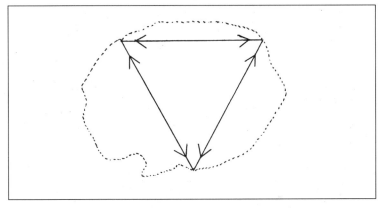

Figure 4

Ideology has many definitions. For the present, I am using only one, abbreviating a concept that I have developed elsewhere (Shinn 1991, Chapter 12). Ideology is a view of society and the world that guides action. As such, ideology is necessary for purposive action. The American Declaration of Independence, the platforms of political parties, the Charter of the United Nations, and START (Strategic Arms Reduction Talks) are all ideological. But ideology, though necessary, is often pernicious, because the partisan interests of individuals and groups shape their ideologies. In fact, ideology has often been defined as the distortion of reality by groups trying to preserve a position of privilege—a definition drawn from Karl Mannheim, who adapted it from Karl Marx (Mannheim 1929).

If ideology is only a distortion, those who love truth (whether scientists, theologians, philosophers, artists, or the most ordinary citizens) ought to try to get rid of it, even though they will never entirely succeed. I readily grant that ideology does, in fact, involve distortion. That does not mean that I want to get rid of it. Although there is no perfect marital love, I do not want to abolish marriage. Although there is no perfect ideology, I do not want to abolish ideology. I want to be aware of it, on guard against its distortions. The distortions rise out of interests of nation, race, class, gender, professions, and any other social groups in which persons find identity.

But taking ideology as a necessary though flawed context of action, what enters into ideology? It includes everything in the triangle of values, science, and politics—plus a lot more. It is a constellation of information, ideas, purposes, prejudices, loyalties, and emotional tones. It takes shape in the slogans that simplify and confuse public debate: class warfare and classless society, free enterprise and social responsibility, melting pot and ethnic identity, right to life and freedom of choice. The irregular, bulging circle in the diagram is a broken line, because the boundaries of ideology are not impervious. New perceptions, feelings, ideas, and social pressures may slip in, jarring and revising the old ideology. But ideology is powerful.

The power comes from the fact that it is not only a composite of many other powerful forces. It is also the lens or filter through which we perceive reality. It depends, to a great degree, on social location. Just as the profile of a mountain differs as we view it from various directions or fly over it, so the world looks different from differing social perspectives.

Today, more than in some past times, authoritative views

of reality are suspect. To be sure, new authorities and new idolatries constantly arise and claim allegiance. But a "hermeneutic of suspicion" pervades society, even though only intellectuals use the term. All of us suspect that others are trying to put something over on us, and we debunk pretentious claims to authority.

We also realize that poor people, oppressed racial groups, and women experience the world differently from the groups who have usually dominated history. The particularity of a social situation limits possibilities of seeing and understanding the world. But particularly is also an opportunity for insight, for perceptions that shake conventional wisdom and folly. The falsity comes when we confuse our particularity with the one "true" comprehension of things.

The significance of ideology emerges clearly in the debates about the new genetics, especially in international gatherings and in interactions between geneticists and people from many other walks of life. Here scientists encounter the deep suspicions of people who see scientific technology as an instrument of militarism, economic imperialism, and (in the case of genetics, above all) a way of defining "normal" humanity according to the ethnic and cultural prejudices of the most powerful and privileged segments of the human race.

Ideology need not always end in irreconcilable battles. Sometimes we can correct ideological errors by recognition of objective facts (the facts that "kick," in Myrdal's term) or scientific reality. Juan Luis Segundo, the liberation theologian who knows the power of ideology, observes: "But a fact is a fact, no matter who points it out" (Segundo 1984, 281). That impact of fact has often broken the hold of ideology in the history of science. Sometimes, also, we can correct ideology through the purging power of our deepest values or of the faith that undergirds all value commitments. That, also, has often happened.

Roger Lincoln Shinn

But these corrections are difficult because ideology is the lens through which we usually see and appropriate science and faith. This lens enables us to perceive some realities, as it obscures others. For example, a telescopic lens enables us to see distant galaxies, but not the double helix of DNA; a microscope reverses the possibilities. But neither lens creates the galaxies or the double helix, which were there before anybody observed them. Given realities may be jolting enough to revise or upset treasured ideologies. But ideology may still prevail in the applications of knowledge. Thus ideology does not determine that people and mice share many genes. But ideology influences every discussion of the possible uses of the new knowledge, whether in medical practice, employment policies, or screening of people for insurance. We have seen in the preceding chapter how it dominates debates about "the oddity of IQ." Out of an almost boundless arena of data, partisans can pull those that support their ideologies.

Our generation is realizing a new awareness of the influence of gender on ideology. Feminists often make the case that women bring to social issues and to science a set of concerns and sensitivities that have often been drowned out by dominant males. I agree. I would only insist that, at the critical point of verification, biologists like Nancy Wexler and Mary-Claire King assess evidence with a scientific objectivity that is not yoked to gender. I endorse the goal of Helen Longino: "developing an analysis of scientific knowledge that reconciles the objectivity of science with its social and cultural construction" (Longino 1990, ix; cf. 1992).

If correction of ideological limitations is difficult, awareness of that problem may be a step toward correction. Openness to others with differing ideologies can help. But at the moment when we claim to transcend ideology, we are most likely to its prisoners.

Most debates about public policy are ideological debates. And ideological debate is the most confusing of all. It is opportunistic, flitting from one point to another of the triangle, choosing whatever fact, ethical claim, or political slogan is expedient. But confusion is not a call to surrender. It is an opportunity for clarification.

THE ROLE OF RELIGIOUS COMMUNITIES IN PUBLIC POLICY

Early in this chapter I declared my intention to develop a model of the process of decision-making, especially in public policy. I said that I intend the model to be descriptive, but that the description might suggest some prescriptions. The descriptive model is represented in the figure of the triangle within the irregular ring. My argument is that public policies *in general*—whether we like it or not—represent an interaction of these three dynamic forces: human values and faiths, scientific information and concepts, and political activity, all operating within an ideological context. I cannot imagine a public policy, benign or malignant, that does not involve some form of this constellation—although the content of the forces can vary immensely.

The description has already edged into some prescriptions. I have urged the importance of recognizing the qualified autonomy, yet the constant interaction and interdependence of the three points of the triangle. It is important to recognize the distinctions among values, science, and politics, and to call on evidence and reasoning appropriate to each realm. It is important to avoid an imperialism of any one realm that tries to dictate conclusions to other realms or prescribe public policy in disregard of the other realms. It is important to recognize the ideological coloration that, often surreptitiously, pervades all realism. Such recognition will not solve all problems; intense debates will continue. But an awareness of the process may lead to some clarification. It may even contribute to that delicate combination of humility and courage that is necessary for healthy development of policy.

To push the argument one step further, we can look once again into the legitimate role of faith within the body politic. With Max Weber, I have said that purpose and values are expressions of faith. Everybody can and does reject many specific faiths. Yet there is no avoiding some kind of faith. William James, picking up Pascal's language about reasons of the heart, wrote: "Science herself consults her heart when she lays it down that the infinite ascertainment of fact and correction of false belief are the supreme good for man" (James 1896, Part IV). That is an esoteric faith, far from the beliefs of most people, but a faith convincing to an international, intercultural community of scientists—and at least somewhat persuasive to many more people who live on the fringes of that community and appreciate it.

The great religions are less esoteric, less elitist than the scientific communities. They are more bound to particular, contingent histories, although some of them have an international, intercultural appeal. In some historical situations—medieval Christendom, Puritan theocracies, avowedly Islamic nations—governments intentionally acknowledge the claims of a specific religious tradition and work to implement the ethical standards of that religion. Intentionally pluralistic societies, committed to religious liberty and to separation of church and state, have renounced that option. In those societies the religious communities—sometimes out of practical necessity, sometimes out of conviction—have come to affirm the values of pluralism and diversity. What, then, is the role of religious communities in influencing public policy? What in particular is the role of theology, the intellectual expression of faith? (*Theology* is a term more familiar in the Christian tradition than in others, but all religious traditions have their theologies or functional equivalents. All the traditions expect people to think.)

SCIENCE, ETHICS, AND FAITH, IN PUBLIC POLICY

First, the religious communities accept and often welcome some restraints. They know that neither their faith nor their theology will, of itself, establish public policies. Particularly in areas like genetics, which open new possibilities and require new decisions, theology must give attention to science. Always it must pay attention to the political process, asking not only what is practically possible but what is desirable to legislate.

But, second, religious communities look for opportunities to persuade. They are parts of the body politic, sharing the rights of all people to advocate their ethical convictions without claiming any monopoly of such rights. Religious communities will not passively accept the opinions of scientists about the uses of their new powers. Knowing the ambivalence of all power, they—and particularly their theologians—will actively interrogate scientists about the human sensitivities and judgments involved in their definitions of health and disease. They will be alert to the unintended consequences of acts undertaken with "good" purposes. They will not, for example, accept the reasoning of Herman J. Muller, who advocated the scientific steering of the direction of biological evolution and threw out the challenging question: "Have we not eventually utilized, for better or worse, all materials, processes, and powers over which we could gain some mastery?" (Muller 1967, 521).

This is not to say that most geneticists agree with Muller or that theologians would, even if they could, automatically veto his proposal. It is to say that the use and direction of new forms of "mastery" is exactly the issue that should be discussed, with participation of people from many vocations and social circumstances. Alert to the temptations of sin and hubris (as the Hebrews and Greeks respectively called it), they will ask about the perils of a self-confident attempt to redirect the processes of nature—

without renouncing the human freedom and responsibility that are always modifying nature.

Theology claims some competence in the criticism of ideologies. That is part of its stock in trade—although it is not itself exempt from ideological distortions. Theologians, with special attention to the ideological biases often hidden in theology, can seek to ferret out and disclose the biases of society. As partners in an international and intercultural community of faith, they have a special responsibility to recognize ideologies of nation, ethnicity, class, religion, and gender. As heirs of a long heritage, they have an opportunity to listen to and speak for the oppressed and those too often despised by elites—whether military, economic, or intellectual.

Any theology draws insights from a particular historical heritage as well as from human experience in general. It is always asking how to relate the particular to the general. By this time in history, most theologians have learned to restrain the natural human inclination to impose their particularities upon everybody else. But they need not be embarrassed that their insights come out of a concrete, and therefore particular history—just as scientists are not embarrassed that modern genetic insights emerged in a particular historical situation.

Theologians, entering into a democratic process of setting public policy, can bring their convictions to a continuing conversation. The testing of these convictions is more difficult for theology than for science, because many processes of scientific verification have earned an intercultural acceptance (at least among scientists) that is hard to achieve on issues of the ultimate loyalties that evoke faith. Even so, many theological convictions find wide resonance in a pluralistic society: the worth of the individual person together with the realization that persons require community, the value of health and healing activities, the concern for justice and for those who suffer injustice, the treasuring of human freedom along with the protest against its abuse, the warning against self-worship, the conviction that governments exist to serve people, the appreciation but not the divinization of nature. Humankind is now asking what our unprecedented scientific powers mean for the renewal and reconceptualization of traditional values.

It is clear that this is a time for exploration. Explorers enter unmapped territories, but not without help from the past. They often carry compasses, charts of the stars, and skills developed over generations of history. Both tradition and innovation are necessary to meet the opportunities and perils of our new scientific powers.

This chapter has centered on method—particularly on the interactions symbolized by the triangle of values, science, and politics within the irregular circle of ideology. The method, I maintain, operates in the most ordinary decisions of life—whether to move on to a new job, when to punish and when to forgive wrongdoers, how to manage the family budget, how to vote. It has implications for many of the ethical issues of HGP: How shall we evaluate new hopes and new risks? How shall we relate concern for individuals to concern for society. What is it to be human? The next chapter will turn to some of these normative issues.

Chapter Five

HEALING, ENHANCING, AND DISTORTING THE HUMAN

Genetic research aims at improved health. For some scientists it has the sheer joy of discovery, one of the perennial motivations of science. But the goals that persuaded the United States Congress to endorse and fund HGP, as we have already noticed (Chapter 2), were the prevention and healing of diseases.

But then the question becomes: What is health? On some days that question is too stupid to waste any time on. We know that we are sick or are getting better. But on other days the question is perplexing. It leads to angry controversies.

We can begin by a distinction in ordinary language. Drugs contribute to the *healing*, the *enhancing*, and the *distorting* of persons. The use of pharmaceuticals for *healing* is usually welcomed. Antibiotics save lives and restore people to health—although warnings against their indiscriminate use show the problematic character of most good things. *Enhancement* is a more ambivalent process. Most people make at least a passing effort to improve their health through diet and exercise, and perhaps a few pills. But as enhancement edges into *distortion*, it becomes objectionable. The use of steroids by athletes—whether human or equine—is forbidden, partly because it upsets the terms of fair competition, partly because it violates basic notions of what it is to be a healthy person or horse. The imperiling of human nature by addictive drugs is an evil—an evil with great attraction for some people and great profit for others. The issues in the use of drugs are magnified in the case of genetics. To investigate these issues requires an examination both of scientific evidence and of human purposes, with a constant awareness of the influence, conscious or unconscious, of ideology.

THE DISTURBING QUESTION OF NORMS

Lurking in all genetic theory and practice is the disturbing question of norms. Terms like disease, ailment, defect, liability, and anomaly all imply some departure from a state of health regarded as normal. Some genetically-based afflictions are so severe that we need not argue about their undesirability. Nobody wants, for self or for offspring, cystic fibrosis, Tay-Sachs, or Huntington's disease. These are *abnormal*; they depart from our standards of normal health. Even with such liabilities, we must be careful that our judgment about the ailment does not provoke injustice toward those who have it—or become the basis for extended judgments about human superiority and inferiority. And to expand the list very much is to move into debatable areas. Soon we are asking perennial questions that have haunted prophets, artists, and philosophers through the ages: What is it to be human? What is good and bad?

James Gustafson raised these questions in a paper entitled "Genetic Engineering and the Normative View of the Human." The time was December, 1969, sixteen years after the discovery of the double helix, nineteen years before the formal launching of the effort to map the human genome. The occasion was a three-day symposium on "The Identity and Dignity of Man," co-sponsored

HEALING, ENHANCING, AND DISTORTING THE HUMAN

by the Boston University School of Theology and the American Association for the Advancement of Science, as part of the annual meeting of the AAAS in Boston. Gustafson asked, "What is the *normatively* human?" and said plainly that he did so with "an evaluative concern." He observed, wisely I think: "The difficulties in coming to a consensus on the normatively human are almost insuperable, yet it is my deepening conviction that some efforts must be made to overcome them" (Gustafson 1973, 46). He did not formulate an answer to his question. Rather, he proposed it for serious thought and made suggestions for methods in the search for an answer.

Even this modest proposal provoked objections. The distinguished philosopher Hans Jonas saw "terrifying" possibilities of the "standardizing or homogenizing of mankind," of creating "monotony and fixations." Gustafson replied by clearly rejecting all grandiose genetic projects or any "single, fixed norm." He advocated a "plurality of values." But he argued that, "given certain kinds of power and capabilities that exist," we had better give attention to the question, rather than leave it to "chance," i.e., to the "people who have the power" to influence the direction of the genetic future (Gustafson 1973, 229–30).

The controversy continues. Cell biologist Fred Ledley sees dangers in the effort to define a normal human genome: "It is dangerous because the notion that there could be a normative human genome also implies that deviations from the normal sequence would be abnormal or undesirable. This would reinforce the tendency to stereotype individuals in terms of characteristics which deviated from the ideal. In fact, no human being matches this ideal" (Ledley 1993). I want to take none of the sting out of that statement, even though I think the question of norms is inescapable.

My purpose is certainly not to frame a definition of normality, then apply it deductively to specific situations. But the question of norms refuses to be suppressed. It lurks in the underground of all genetic discussions, occasionally erupting into conscious deliberation. By investigating the actual normative judgments that operate in all research and discussion, we can raise hidden norms to scrutiny and criticism.

Warnings

History gives some pointed warnings against prescriptive norms. These are of two principal types.

1. Skewed norms

Individuals and societies, moved by an invincible impulse, treasure their own normality and seek to impose it on the world at large. Afflicted with self-regard, they universalize their particularities.

The Greek historian Herodotus, in the fifth century B.C., observed this common human foible. Known as "the father of history," he might also be called the father of cultural anthropology. He traveled widely and sprinkled his pages with extended reports, sometimes direct observation and sometimes hearsay, of the customs of many peoples. He concluded: "For if it were proposed to all nations to choose which seemed best of all customs, each, after examination made, would place its own first; so well is each persuaded that its own are by far the best." As an example, he told how Greeks were horrified at the custom of the Callatian Indians who ate the bodies of their fathers, while the Callatians were equally shocked that the Greeks would burn the bodies of the dead. Then Herodotus referred to Pindar's poetic saying "that use and wont is lord of all" (Herodotus 1921, III, p. 51).

Plato is usually regarded as more profound and sophisticated then Herodotus. He tried his best to transcend the relativism of Herodotus and the Sophists by appealing to transcendent values and norms. This was an admirable aim, but Plato never guessed how ideologically tainted his values were. In *The Republic* he proposed a hierarchical society, dominated by an Athenian elite, who would not only govern but would deceptively manipulate the society into a form of genetic planning expressing their own values. In his "fatal provincialism," writes Herbert J. Muller, Plato "wrote as if Herodotus had never written" (Muller 1952, 120).

The issue lingered through generations of history, then erupted in the seventeenth century. Pascal, with characteristic wit, pointed out that the normal is usually simply the familiar. "Who is unhappy at having only one mouth? And who is not unhappy at having only one eye? Probably no man ever ventured to mourn at not having three eyes" (Pascal 1670, Fragment 109).

"Custom is our nature," he wrote. "What are our natural principles but principles of custom?" "As custom determines what is agreeable, so also does it determine justice" (Fragments 89, 92, 309; cf. Fr. 97, 252, 325). Geography and climate determine our norms. "Three degrees of latitude reverse all jurisprudence; a meridian decides the truth. . . . Truth on this side of the Pyrenees, error on the other side" (Fr. 294). Pascal understood very well that his relativism teetered on the edge of nihilism. He had his ways of avoiding that nihilism by a wager of faith; but the immediate point of his relativism was a drastic warning against imposing customary cultural norms on people with different cultural norms.

In more recent times, advocates of eugenic practices have reinforced the importance of Pascal's warnings. Nazism is the horrendous example that haunts all contemporary discussions of genetic practices. It is encouraging to notice that scientific genetics often refutes the distorted genetics used by Nazis and other racists. Their vicious errors do not invalidate more humane efforts to map the human genome and apply the knowledge to human health. But the imperious Nazi doctrines (like the doctrines of Plato long before) are a warning against the insidious distortion that creeps into efforts to define normalcy.

The warning is the more pertinent when we realize that many Americans of the nineteenth and twentieth centuries, seeing themselves as idealistic humanitarians, made comparable errors of projecting their self-image upon the world as a standard of normalcy. They did not try to annihilate deviants from their own norms, but they took their idiosyncratic preferences to be the "right" norms for a diverse world (Kevles and Hood 1992, Chapter 1). To point to the racism and sexism in many of the "liberal" moralists of the past is not to gloat over our superiority. It is to recognize the need for self-examination, with the hope of discovering our own limitations before some future generation unveils our silliness.

Contemporary sociologists and psychologists update the insights of Herodotus. Peter Berger and Thomas Luckmann, whom I have mentioned earlier, show how every society tries to maintain a symbolic universe "in the face of chaos." Therapy, "a global social phenomenon," becomes a form of "social control" that isolates deviants and dissenters who challenge the prevailing norms. An "esoteric subuniverse of medicine . . . shrouds itself in the age-old symbols of power and mystery" in order to maintain its authority. It is the more effective because it seems to be benign, and it is frequently internalized by deviants so that they cooperate in their own marginalization or suppression (Berger and Luckmann 1967, 88, 103, 112–14). In a more radical way, psychologist Thomas Szasz, in a series of twenty-three books, has sustained a polemic against the psychiatric practices that function to get unwanted persons out of the way (See esp. Szasz 1961 and 1994).

These arguments can be carried to foolish extremes, as I think Szasz does. But his objections are useful reminders that all therapy presupposes some idea of normality, which itself needs critical investigation.

A telling example is the record of changing understandings of homosexuality. These vary greatly throughout history, but three are instructive. The first regards homosexuality as a moral perversity—a dominant attitude in much modern Western history. The second, making no moral judgment, regards homosexuality as an illness. This purports to be a more benign attitude, but many gay and lesbian people protest that they find it equally offensive. The third regards homosexuality as one form of normality. In 1973 the American Psychiatric Association removed homosexuality from its list of mental disorders, thus tacitly including it within the range of normality. All three attitudes embody some norms, explicit or implicit, of human personality. Any inquiry into genetic influences on homosexuality provokes heated controversies, as I have noted in Chapter 2.

In terms of the logic of this book, much of the scientific debate should be settled on scientific grounds. But the public appropriation of the findings will involve the purposes and evaluations of individuals and society, with all the perils and promises inherent in human norms.

2. Illusions of perfection

In addition to skewed norms, a second kind of problem arises in the search for the normative human. It is the illusion of perfection. It is the worse because the illusory perfection is generally defined in skewed terms. But even if it could ever be pure as the driven snow, it would have its perils.

The illusion is evident in the desire of parents for a "normal child." By a normal child, they usually mean a child free of genetic and congenital liabilities. The trouble with this is that "normal" people generally have a few genetic liabilities.

Roger Lincoln Shinn

Yet there are genetically-induced liabilities that are so devastating that nobody would desire them for self or offspring. I have earlier pointed to the possibilities, in the case of some ailments, of genetic testing of parents and of fetus, and the sober choices it affords. But no science, now practiced or on any horizon, can detect all genetic aberrations, let alone cure them. We human beings can, should, and will reduce some perils to life and health. But nothing in the progress of genetics, spectacular though that has been, annuls the insight inscribed in the myths, poetry, and religions of many cultures, that human life is finite and vulnerable. Perfection is a deception. It is normal for persons to have some abnormalities.

I have been describing two types of warnings against defining a normative human ideal: the universalizing of peculiarities to the disparagement of the particularities of other people, and the illusion of perfection. The warnings are serious. I do not intend to forget them. But they are not prohibitions. They do not annul all differences between health and sickness. It remains the case that all discussions of genetic practice, including the warnings themselves, include some implicit affirming of norms or groping for them.

THE INTERACTION OF BIOLOGY AND CULTURE

The search for norms is complicated by the intricate interaction of biology and culture in all human life. Here I return to the earlier discussion of nature and nurture, now in a different context. Sometimes we can readily distinguish biology and culture. I and all who read this book are biological beings; we must eat to live; we are going to die; we have genomes, partly shared with one another and partly unique, with twenty-three pairs of chromosomes. We are also cultural beings: I communicate, more or less successfully, in a shared language. I report genetic information

Healing, Enhancing, and Distorting the Human

unknown through most of history. I assume a "modern" or "postmodern" consciousness.

But the biological and cultural, though distinguishable, are not separable. The conventional arguments about nature and nurture, we have seen, are fumbling attempts to assess the relative importance of each, and they never come to a firm conclusion. More interesting is the intricate relationship between the two.

A few characteristics of physical health, in its most elemental terms, are predominantly biological. Norman human body temperature is, for the most part, a statistical average—even though the measurement of temperature is a cultural achievement. Nobody is insulted by news of a subnormal temperature or of fever. On greater issues, I cannot imagine any culture in which it is desirable to live with congested lungs or to suffer disintegration of the central nervous system in childhood.

But culture quickly enters into questions of physical normalcy. Some, certainly not all, limitations of eyesight that were once serious liabilities, have become trivial because of corrective lenses. Smallpox, once a scourge of humankind, has—we think—been eradicated; and many once-fearful diseases are easily preventable through vaccines, a cultural achievement. New perils, like AIDS, are unquestionably biological, but their spread through populations is quickened or slowed by cultural practices.

Pigmentation of skin and hair is a biological fact, largely genetic. But what is normative? Biology tells us a little: blonds and blondes have a hard time of it on tropical deserts. But culture, not biology, has made dark skin a practical liability in many parts of the earth. The Indian caste system and Western slavery and segregation are cultural developments, exploiting skin color as a pretext for domination. Biology (including centuries of natural selection in various climates) accounts for differences in color; culture makes some of those differences invidious.

Roger Lincoln Shinn

Those cultural norms are so pervasive that they may become internalized, even by those who suffer from them. Only a few decades ago, magazines for African-Americans in North America carried advertising for hair-straighteners and complexion-lighteners. We can guess that, if a laboratory had developed a genetically-engineered product for the same purpose, it would have found a market. Today most people see the absurdity of such notions. They were efforts to alter biology to meet cultural prejudices. An easier answer—not yet fully comprehended in our society—is the recognition that "black is beautiful."

Physical size is another biological trait, intricately interwoven with culture. Although diet has some relation to size, genetics is one determining factor. Nobody wants to rival elephants or mice in size; people set human norms, which vary. The Masai and Watutsi (Tutsi) people of Africa are, in general, tall. A height of seven feet is frequently among them. Given the peculiarities of human nature and culture, that height may have something to do with their aristocratic domination over some other African peoples. The Pygmies, by contrast, tend to be short. That is biological fact. It is a curious fact in the self-created world of racial prejudice, because it is associated with definable ethnic groups, but the groups belong to the same black, African "race," as that confusing term is defined in popular talk.

But the traits of height, though biological, have cultural meaning. To "stand tall" is, in our culture, to show courage and pride. The term usually refers to males—another biological classification that has deep, often discriminatory cultural meaning. Words have different meanings within different subcultures. In the National Basketball Association a "small forward" may measure 6'8".

Height has no relation to essential human worth, but it has important functional meanings. A coach, assembling a basketball team, might (quite without racial prejudice) scout Masai rather than Pygmies; somebody looking for jockies might skip by Masai. These examples are egregious uses of categories from my culture, exported into the cultures of other people. I use the examples deliberately, to show the bizarre ways in which people relate culture to biology.

To take an example closer to home, left-handedness is an inconvenience to most people, apart from a few ball players who make an advantage of it. Writing desks in schools, scissors, and many tools are designed mostly for right-handed people. For generations many American parents and teachers tried to "break" the left-handedness of children and force them into right-handed habits, sometimes with harmful effects on personality. The word sinister (derived from the Middle English term for left-handed) is a typical human transfer of physical to moral categories. After a lot of fuss and pain, our culture is learning to adapt to left-handedness rather than force lefties to adapt to culture.

More generally, the genetically-based traits that are most helpful in a high-tech society are not exactly those most helpful in a primeval jungle. Any attempt to reach toward a conception of normative humanity must be wary of the inveterate human tendency to impose particular cultural norms upon all humanity. Looking ahead to further developments in genetic science, we need to ask: When should we try to alter genetic qualities? Past experience urges a caution. A caution is not a total prohibition. Natural selection has, over millennia, made some alterations; and there may be occasions for purposive modifications. But too much human thinking is a mindless extension of cultural peculiarities to criteria for judging all humankind.

When individuals or groups deviate from cultural norms, there are times when they may be persuaded or required to conform to the culture. There are times when genetic planning should, by reduction of diseases, encourage a kind of conformity to norms of health. But more often the culture should engage in self-examination, asking: When should the culture change to meet the protests of those in disadvantages? When should it enlarge its compass to include variations within its membership?

Some Clues to Human Norms

The questions about norms of selfhood will not go away. I have emphasized the valid warnings against skewed and utopian norms. I have argued that the complex relations between biology and culture often lead to malicious or foolish normative judgments. I hope to keep my warnings in mind as I make some ventures toward formulating normative judgments.

But such judgments are inevitable. Week by week the media report new developments in human genetics and controversies about their possible applications. All the controversies assume some normative ideas about human good and bad. These need scrutiny.

The scrutiny is partly scientific. Some data about human beings and their health can be quantified and analyzed. When genetic knowledge informs us of probabilities of life expectancy or definable perils to life, that information enters into our ideas of health and disease, of normality and abnormality. But more than science is required. Our values, purposes, ideologies, and faith enter into our normative assumptions. We cannot avoid questions of philosophy and religion, asked for centuries, in a variety of languages: Is there a normal selfhood, an ego, an essence of humanity?

Everyday language speaks of the human in two quite different ways. In a bare, empirical sense, the human refers to any characteristics of our species, as distinguished from the inanimate, the vegetable, the canine or feline or bovine or equine. In this sense of the term, Genghis Khan and Isaiah, Adolf Hitler and Mahatma

Healing, Enhancing, and Distorting the Human

Gandhi, Saddam Hussein and Mother Theresa are all equally human. But usually we do not talk that way. We speak of inhuman and inhumane behavior, of humanizing and dehumanizing acts and institutions. Here we appeal to qualitative meanings of the human.

My present purpose is not to explore the wealth of meaning in the concept of the human. It is the more limited attempt to identify some clues to human norms as they pertain to genetic exploration and practice. I shall identify six attitudes or insights, and show how they may lead to some guidelines for genetic activity. I say guidelines rather than prescriptions and laws. The guidelines may point toward some prescription. But too many of the current debates leap to prescriptive conclusions.

1. Physical health

I start with the most obvious. Without minimal physical health there cannot be human life. Something more than minimal health is needed to sustain the energies that create families, build economies, organize community life, and develop the arts and sciences that are typical of our race.

The criteria of physical health are, in some respects, objective and obvious. I have given bodily temperature as an example. If it rises or declines many degrees from the measurably "normal," the individual does not survive. Normality, allowing for moderate variations, is close to an arithmetic average. Respiration, blood circulation, and the functioning of a nervous system are objective essentials to life and health. Sometimes the measure of normality is not quite arithmetic. It is statistically normative for people to have two or three colds a year, but those colds are not part of normal good health. Even so, we often know, with a fair degree of objectivity, what we mean by health and sickness.

But health—including a healthy diet and lifestyle—is not

identical in arctic and tropical regions. Since health is in part culturally defined, we should not leap too quickly to universal definitions of health.

With these several qualifications, health and sickness are widely recognizable criteria of the quality of life. The elimination of smallpox and the suppression (never perfect but considerable) of cholera and plague are epochal human achievements. The treatment of genetically-transmitted ailments, sometimes with the help of genetically-engineered drugs, is another achievement.

Two cautions are in order. First, genetic engineering involves special risks, which call for special cautions. Some genetic proposals, if they go wrong, can cause disaster, an issue that requires more attention in the next chapter. Second, health is always tentative and precarious. Human life is fragile, and it moves toward death. Arrogant gestures, forgetful of our finitude, are foolish. They can do harm.

Proposed guideline: Health is a good. Healing is a good. Genetic investigation and practice, directed toward healing, are beneficent—provided they guard against excessive risk, rash denials of human frailty, and partisan definitions of normality.

2. Humankind and nature

Humanity is inalienably part of nature, yet a very special part with the ability to investigate nature, love or resist it, exercise some precarious control over nature, whether in ourselves or in our environment. The new genetic knowledge heightens the paradox of our relation to nature. On the one hand, it shows our close relation to chimpanzees, mice, and bacteria. On the other hand, it shows us how to rearrange the genes in other forms of life and in ourselves, as though we were managers of nature.

Throughout history nature has evoked awe and wonder,

Healing, Enhancing, and Distorting the Human

often reverence. Presumably everything that we know as nature follows from a Big Bang aeons ago. To ask about anything before the Big Bang is an unanswerable, some would say a meaningless, question. Try to find words to describe that unique event: dramatic, spectacular, marvelous, tremendous, and cataclysmic are far too pale. In it, somehow, were all the possibilities that have unfolded since. Nobody really knows how that cosmic event led to this beautiful planet, its many forms of life, its human species with forty-six chromosomes, its Beethoven and its Shakespeare, its Buddha and its Christ, its Einstein and its Heisenberg, its biologists who explore the human genome.

This nature both supports us and imperils us. Apart from it we could not exist. Yet it frustrates, threatens, and sometimes torments us. Sooner or later, it kills us. So our awe for nature must be suspicious, skeptical, ambivalent. Nature is not just O.K. Nature, despite the persuasion of some ecologists, does not "know best." In the biblical tradition, God looked at the creation and called it "very good." But even the original idyllic garden needed a human caretaker. And a serpent was lurking in the greenery. Nature has its destructive side—destructive, certainly, of human values and of any trans-human values that we can imagine. Appreciation—respect, awe, love—of nature had better be tinged with apprehension.

This guarded appreciation is the context for any evaluation of genetic practice. Joseph Fletcher, suspicious of nature and its processes of "genetic roulette" (Fletcher 1974), favored scientific interventions to improve on nature. Erwin Chargaff had greater respect for "the evolutionary wisdom of millions of years" (Chargaff 1978, 190). My own judgment leans toward Chargaff's respect for nature, touched with Fletcher's suspicion.

Margaret Mead, thinking in 1949 about the world of high technologies that were soon to include genetic techniques, asked:

Roger Lincoln Shinn

Was it possible that modern man might forget his relationship with the rest of the natural world to such a degree that he separated himself from his own pulse-beat, wrote poetry only in tune with machines, and was irrevocably cut off from his own heart? In their new-found preoccupation with power over the natural world, might men so forget God that they would build a barrier against the wisdom of the past that no one could penetrate?" (Mead 1995, 19).

The very language of genetic engineering intensifies the poignancy of those questions. We cannot engineer spiral nebulae, showers of meteors, or even sunsets. Yet we, so dependent on our genes, are beginning to engineer ourselves. We had better do so with fear and trembling.

Proposed guideline: Nature and its awesome, intricate ecosystems deserve our profound but guarded respect. Nature does not dictate norms. We can ask of every genetic intervention in nature whether a modification of culture would be preferable. But we can intervene, with due caution, to protect significant values.

3. Freedom-in-community

I have earlier (Chapter 3) had my say about freedom as a human reality in the midst of biological and social determinisms that are very real but not omnipotent. Here I look at it as a norm of selfhood.

Freedom is one of the great words of the American culture. Schoolchildren learn to sing, "From every mountainside, let freedom ring." They memorize sayings like, "Give me liberty or give me death." If in our culture we were to find one word to signify the qualitatively human, it might be freedom. But many societies, past and present, have put a higher value on order, on security, sometimes on harmony. Some order and security are a necessity for life, some a necessity for a good life. The human being, with its

HEALING, ENHANCING, AND DISTORTING THE HUMAN

prolonged infancy, needs community from the beginning for survival. Even the most self-reliant adults rarely test the possibilities of existence outside community. Although community restricts some expressions of individualistic freedom, it expands opportunities and therefore freedoms. So I use the term *freedom-in-community* as a norm of human nature.

Genetic practice can enhance or imperil freedom-in-community. There is no gene for freedom that can be inserted into people bound by necessity; to manipulate people into freedom is a contradiction in terms. But to the extent that genetic skills overcome destructive diseases, they free people from burdens that are a kind of slavery. To the extent that those same skills turn persons into manipulated mechanisms, they destroy freedom.

Yet manipulation is not just a dirty word. There are mechanisms in the self that can be manipulated. Sometimes I crave manipulation: I want a pharmaceutical prescription for an illness, I want an ice pack on a sore muscle, I want surgery to correct a malfunction (whether genetic in origin or caused by an accident). Sometimes I manipulate myself, as in a regimen of exercises. But it is I who wants to be manipulated, who wants to manipulate. That is why "informed consent" is one of the basic ethical rules of medical practice. Nobody asks for the informed consent of the double helix and its nucleotides, but people demand informed consent for themselves. Even informed consent cannot be an absolute rule or a possibility at all times—a topic for Chapter 6. But inherent in selfhood is the craving for freedom and self-determination. That leads to the deep suspicion and resentment often felt against any elite that appears ready to manipulate persons, whether by genetics or drugs or behavioral conditioning, for the goals of the manipulators.

Vaclav Havel, the Czech playwright and first president of post-Communist Czechoslovakia, put it this way: "People are not just racks on which to hang various organs—kidneys, stomach, and so on—that can be repaired by specialists, as you would repair a car. They are integral beings in whom every part is intimately interrelated, and in whom everything is mysteriously connected to the spirit" (Havel 1992, 13). All the marvels of biochemistry and genetic knowledge do not dispel that mystery of the spirit. And a norm of that spirit is freedom-in-community.

Proposed guideline: Freedom-in-community is one of the normative marks of humanity. An ethically responsible program of genetic research and practice, while recognizing that the human body includes causal mechanisms, some of them genetic, will remember that persons are more than mechanisms or collections of mechanisms. It will recognize the mystery of selfhood and will seek to protect freedom-in-community.

4. Diversity

Diversity is one of the marks of human life. The individual diversity of persons is far greater than that of other species. A superficial evidence is taste in food. The food preferences of dogs, cats, deer, horses, and cattle are fairly predictable. But gourmet restaurants offer a great variety of choices. Children differ from their parents, and siblings differ from one another, sometimes causing family arguments. The range of diversity in personal ambition, artistic creativity, music, the visual arts, romantic attraction, athletic preferences, vocational aptitudes, and friendships is immense.

One of the greatest fears accompanying modern genetics is that somebody or some group will define normality so as to exclude many people and perhaps reduce the human race to monotony. I have already criticized the tendency of races and cultural groups to act as though their particularities were universal norms for humanity.

HEALING, ENHANCING, AND DISTORTING THE HUMAN

Almost all the discussions of ethics and genetics emphasize the value of diversity. One reason for this is biological. Edward O. Wilson writes: "Species diversity—the world's available gene pool—is one of our planet's most important and irreplaceable resources" (Wilson 1989, 114). Human diversity also has survival value. Throughout history people have lived in many regions, climates, and cultures. The best adaptation to any one situation may not be a helpful adaptation to far different situations. Eskimos would have difficulty surviving in the tropics; the people of sub-Saharan Africa would not easily manage the transition to Newfoundland. Most of the computer addicts of Southern California lack survival skills for life in a jungle or savannah populated by lions and elephants. The most successful dwellers in a hunting-gathering society would have troubles in the asphalt jungles of dense cities. To a great extent, the differences are cultural; people have an ability to adapt. But to the extent that the differences are genetic, diversity in the gene pool is important to the survival of the human race.

Despite the wishful thinking of people who assume that the achievements of their own culture are permanent, none of us can really predict the distant future. The possibilities are startling. High energy consumption might heat up the planet. A nuclear war might devastate civilization. A cosmic projectile might slam into the earth (like the one that apparently doomed the dinosaurs). Robots might reduce the quantity of physical labor and turn us all into computer wizards, or energy shortages might reinstate the importance of muscular labor. In all cases, the future may value genetic traits not now much esteemed. Genetic diversity is a kind of insurance against the unexpected.

A second reason for esteeming diversity is the sheer joy of the rich variety of human expression. A world without symphony orchestras would be poorer for the loss, but not as poor as a world in which the only music was symphonies. Modern artists (Picasso is an example) have found inspiration in art once derided as "primitive."

A third reason, more strictly ethical, is the humble recognition of the values of peoples with different gifts from our own. Contempt for others—for people of other races, religions, and cultures—ends in arrogance or self-contempt.

Most of these differences are primarily cultural. But to the extend that there is a genetic basis for them—we seldom know for certain what that extent is—genetic diversity is an asset to us all.

However, diversity, like other good things, must not become a mindless cliché. It is not an unlimited, unqualified value. Consider an example. It is possible to irradiate seeds of plants to increase the rate of mutations and consequent diversity. Most of the mutant plants are inferior to the originals, and go on the compost pile. But a few out of a thousand may please the experimenter, be propagated, and find a market. Nobody complains. The experiment can be declared a success. A similar experiment can be done with fruit flies—although the result is not likely to be marketable. A proposal to irradiate the germ cells of dogs would arouse public protest. And a plan to do the same with persons would horrify the populace. Again we confront the mystery of human dignity. Diversifying can come at too great a cost.

Moreover, there are diseases that nobody would deliberately perpetuate for the sake of human diversity. Natural selection has worked through the ages to reduce some genetically-transmitted ailments. Modern medicine sometimes finds treatments that heal diseases or make afflictions less burdensome, thus annulling the effects of natural selection. If such healing reduces suffering and opens opportunity to diverse human beings, it is desirable. But it is not desirable deliberately to perpetuate genetic liabilities for the sake of human diversity. I shall return to this issue in the next chapter.

Proposed guideline: Diversity is an asset to be treasured.

That does not mean that norms for human life are so particular and so varied that there can be no general norms. But the greater error, throughout most of history, has been the scorn and cruelty directed toward those who deviate from the norms of the dominant and powerful. Any general norms should include diversity. Attempts to normalize deviant types should give a strong weight to the uncoerced choice of those to be normalized.

5. Imagination and reason

I am avoiding the word intelligence, because I am not here concerned with the oddity of IQ. I am concerned with a unique combination of human gifts. Although imagination and reason are often set in contrast, they belong together. They are not strictly quantifiable. Imagination is the ability to see what is not obvious or to envision possibilities not yet realized. It has recently been disparaged as "the vision thing," but it is essential to creative art, science, and statecraft. Rationality is the ability to verify conjectures, combine particular insights into larger constellations of meaning, and figure ways of getting to goals. It is critical thinking, including self-criticism.

So far as we know, imagination and reason are distinctively human traits, although we do not know what goes on in the heads of dolphins or foxes. We do know that various species have various ways of coping with reality. Among medical people there is a saying that the viruses and bacilli have an advantage over us, because they are so much older and wiser. Their wisdom is not that of the scientist mapping their genome or ours. It is closer to "street smarts," but is not that either. These forms of life have an ability to mutate and to reproduce at a tremendous speed. About the time when our antibiotics seem to do them in, somewhere on earth a new mutant resists the antibiotic and multiplies to menace us again.

Instinct is another ability, weak in humankind but amazingly competent in insects and animals. Think of the organization of a colony of bees or ants. Or think of the migration of birds, some of whom travel thousands of miles twice each year, finding their way with an inborn skill unknowable to human beings.

Human imagination and reason are different. Our tactic for survival is not "mutate and multiply." Nor is it to perform tremendous feats of the instincts, of which we have few. Instead, we have an ability to improvise, meet our environment, develop our values, and pass on our achievements to future generations, who will revise what they learn from us and try new ways.

Our history is largely a history of culture. Our advantage over our ancestors 10,000 years ago is not in brain or brawn; we have about the same genetic abilities they had, perhaps weakened a little because we have softened the cruelty of natural selection. Our advantage over those distant ancestors is that we are the heirs of the cultures that have produced agriculture, the arts and sciences, the forms of human organization that we call nations and civilizations.

But it is we (the human race) who have done that—not the ants and tigers and eagles. There is a genetic base to imagination and rationality. We did not invent or produce it. We may call it a gift of God or of nature; we must acknowledge it as a gift. We can cherish and guard it. If we want to enhance human life, culture will be our main means. Bernard Davis spoke for most biological scientists (if not for the reports about them in the tabloids) when he said: "I doubt that advances in molecular genetics will have any impact, in the foreseeable future, on intelligence and personality in man" (Davis 1973, 31).

So imagination and reason are normative marks of human life. They are desirable. That does not mean that we can rank persons with a thermometer measuring imagination and reason. The wisdom of the principle, "one person, one vote," is the recognition of a basic human worth, independent of our prejudiced guesses as to who has the livelier imagination or the wiser reason. In elections all of us can try to choose leaders with exceptional abilities, not always with notable success.

In valuing imagination and reason, we can regret those genetic ailments that mar imagination and reason. We hope that our children will escape them. We do not love our children in proportion to their abilities of mind and spirit, but we welcome any social, medical, or genetic achievements that reduce their afflictions.

Proposed guideline: Imagination and reason are gifts to humanity, to be valued. They do not determine human worth. But when they are threatened by genetic ailments, healing is welcome.

6. Character

Here we come to the most puzzling and mysterious of all human qualities. What we mean by character certainly has a genetic base. It is not present in any biological species except the human. There are analogies—we do not know how close—in animals. People speak of faithfulness and affection in animals. Loren Eiseley writes eloquently of the hawk that can "cry out with joy" and "dance in the air with the fierce passion of a bird" (Eiseley 1956, 193). But human character—with all its richness, its perversity, its contradictions, its struggles against temptation—is unique.

One of its gifts is the capacity to love. The self is incomplete without a devotion to persons and ideals beyond itself. Love is often cramped and distorted into partisan loyalties that embitter life. But even then it is a sign of a craving that breaks the boundaries of isolated individualism.

Although there is a genetic relationship, the connection between the genome and character is hard to trace. J.B. Scott has pointed to experimental evidence of "the lack of congruence between genes and complex behavioral characters" (Hoagland and Burhoe, 1962, 8). J.B.S. Haldane says that "we know the genetic basis of few desirable characters" (Haldane 1963, 345).

The issue emerges with fascinating allure in the writings of Herman Muller, whom I have mentioned twice before. He was ambitious to improve the human race genetically. Since gene-splicing was not a possibility then and he thought it an unlikely prospect, he urged a vast program of artificial insemination, using the sperm of men chosen for their genetic prowess. Small-scale experiments have come out of his advice, but the idea never became widely popular—for good reasons, I believe. But my interest at this point is Muller's dilemma on the issue of character.

Muller knew what human qualities he wanted to enhance. He was too idealistic to settle for physical health and intelligence alone. In fact, his first aim, deliberately put up front, was improvement of character. The qualities he wanted were an increase of "brotherly love" and "deep and broad warmheartedness." But he did not mean these in any timid sense. He also wanted initiative and aggressiveness in the sense of "independence of judgment and moral courage" (Muller 1967, 528–29).

Muller's critics immediately pounced on his assumption of a close correlation between genes and traits of character. Assuming that there are genetic sources of aggression and passivity, how do these relate to moral courage and love? Would we, for the sake of world peace, like to reduce human aggression? Perhaps. But would we have wanted Mahatma Gandhi and Martin Luther King, Jr. to have been less aggressive? Would not a ruling elite want the population-at-large to be less aggressive? The issue of character is not limited to elemental traits transmitted by DNA; it is a matter of what persons and communities—in their freedom, imagination, and reason—do with their hormones and impulses.

As Reinhold Niebuhr put it, "Human consciousness not

only transcends natural process but it transcends itself. It therefore gains the possibility of those endless variations and elaborations of human capacities which characterize human existence. Every impulse of nature in man can be modified, extended, repressed and combined with other impulses in countless variations" (R. Niebuhr 1941, 55).

Muller's proposals were, in a real sense, elaborations of Plato's ancient ideas—on a far greater scale made possible by frozen semen kept in sperm banks. One moral advantage over Plato was that Muller intended the plan to be voluntary, without either the coercion or deception that Plato (satirically, some have suggested) advocated. But he still had to face the problem of Plato: Who will guard the guardians that run the plan? For Plato, the guardians were to be trusted to do what was best for the society. Muller had in mind a more democratic process—guided, of course, by a scientific elite. Yet he was shrewd enough to see the weaknesses in his own ideas. What reason was there to think that the operation would select donors who shared Muller's values of love and courage? Muller, a social radical who for a time admired Lenin, thought that contemporary Western culture was so corrupt that it could not be counted on to select donors whom he would approve. In particular, he worried about the possibility of a "genetic race," comparable to an arms race.

More recently, as genetic engineering has advanced beyond Muller's expectations, Ruth Hubbard and Elijah Wald have sounded the alarm. They recognize that the power of scientific knowledge is likely to be misused. They have become fierce critics, not of all genetic knowledge, but of the uses of that knowledge by those who are likely to control it (Hubbard and Wald 1993).

"Knowledge is power," wrote Francis Bacon four centuries ago. Since then, history has magnified the import of his claim. The purposes of those who use the power and the shape of the institutions within which they work will determine whether the power of genetic knowledge will liberate or oppress people.

So we return to the question that began this chapter. Are there norms of humanity? Is there a human dignity that resists tampering, that is morally inviolable? I have answered that there is no self that floats serenely above history, untouched by the chemistry of nucleotides and amino acids, unswayed by the movements of history and culture. But I have also said that the self is not the plaything of necessity and chance. That self, dependent on genetic processes and culture, has an ability to direct the processes that sustain its life.

In the biblical tradition the human person is created of the dust of the earth, in the image of God. A secular echo of that belief uses the language of humanization and dehumanization. Character makes a difference. Human dignity does not depend on brilliance or moral perfection. All of us are flawed. One theological tradition refers to our "alien dignity"—a dignity not achieved or earned, but the gift of a Creator, ingrained in every person. We dare not despise it in anybody, lest we despise it in ourselves.

Proposed guideline: Genetic knowledge is intimately linked to human character, but character (social and personal) will direct the uses of the knowledge. Both the most grandiose expectations and the indifference to genetics are mistaken. We cannot entrust genetic power to any elite—military, financial, or scientific—without violating the gift of human dignity. The users of genetic knowledge may, or may not, make it a blessing to humankind.

The six proposed guidelines will be disappointing to anyone who—like me, in some of my moods—wants specific directions for the application of genetic power. None of them can be legislated or inscribed in precise behavioral codes. They refer less to precise decisions than to the personal and cultural climate in which decisions are made. I come back to Whitehead's theme of "the secret imaginative background" of all our policy decisions. What this might mean for a more specific set of decisions is a theme of the next chapter.

Chapter Six

A CRUCIAL ISSUE: MODIFICATION OF GERMLINE CELLS

The arc of this study began with attention to scientific discoveries that require ethical decisions. It then moved through three general issues that emerge from the new knowledge and power: the meaning and reality of human freedom; the relations among values and purposes, science, and political life; the problem of defining health. Now it returns to a specific issue that excites and troubles us today: the issue of the possible modification of human germline cells. I flagged this in Chapter 2 and promised a return to it.

    This issue is not so immediate as some others that people are meeting daily: Should a couple planning marriage ask for genetic testing, with the possibility of breaking off their relationship or resolving on a childless marriage? Should parents ask for a prenatal diagnosis with the possibility of aborting a fetus? Should a youth, with a family history of a mid-life disease, ask for genetic testing, even though no treatment is known for the suspected ailment? I am selecting for attention a more remote—possibly only a little more remote—issue. I select it because the discussions have

Roger Lincoln Shinn

already heated up and because so many important questions come to a focus here.

THE ISSUE AND THE SPECTRUM OF OPINIONS

To recall a distinction from Chapter 2, genetic therapy for somatic cells is already in experimental stages. It offers hope for the healing of some disabilities in an individual, but does not hinder that person from passing on the ailment to his or her children. Likewise, it is the patient, not the patient's genetic heirs, who accepts the risks involved in the treatment. Germline therapy might mean far more: the elimination of the ailment in a fertilized egg and all the future generations that follow from that egg throughout the centuries. We must examine the debate and the issues involved. Figure 5 represents roughly the spectrum of opinions.

1. Emphatic ↔ 2. Cautious ↔ 3. Restrained ↔ 4. Enthusiastic
   rejection   openness   expectation   anticipation

Figure 5

There are no set barriers between the positions, because at the fringes each merges into its neighbor(s). Advocates of one position may move to another in the light of new information. But it is possible to identify the four positions and point to representatives of each. I can here give a few samples, before investigating the reasoning behind them.

1. Emphatic rejection.

This position is public policy in some countries. The Parliament of the Council of Europe in 1989 called for a prohibition of germline research, but later allowed exceptions for preven-

A Crucial Issue: Modification of Germline Cells

tive or therapeutic purposes. The German Parliament in 1990 prohibited interventions in the human germline. Switzerland, after a public vote, enacted a constitutional ban in 1992.

Some nongovernmental groups take equally strong positions. In the United States the Council for Responsible Genetics "strongly opposes the use of germ line modification in humans" (Council for Responsible Genetics 1992). *The Book of Discipline of the Methodist Church* says: "Because its long-term effects are uncertain, we oppose therapy that results in changes that can be passed to offspring (germ-line therapy)" (United Methodist Church 1992, 97–98).

In June, 1983, Jeremy Rifkin, a prominent social activist and crusader, released a resolution opposing all "efforts to engineer specific genetic traits into the germline of the human species." His accompanying interpretation included "a call upon Congress to prohibit genetic engineering of the human germline cells" (Rifkin, 1983a). The stand of Rifkin, who in the same year published *Algeny* (Rifkin, 1983b), was not surprising. What was surprising was that he won the endorsement of the chief officers of most major Protestant denominations, several Roman Catholic bishops, and a very few prominent theologians and scientists. Some signers later expressed second thoughts. The participation of the church leaders was unexpected, inasmuch as many of them had earlier approved differing positions of the National Council of Churches in the U.S.A. and the World Council of Churches. That requires us to move to the second position on the spectrum.

2. Cautious openness.

Robert Cook-Deegan expresses this position in his advice: "Keep the window open a crack" (Cook-Deegan 1994, 217–20). He plainly does not expect practice of germline therapy in the early future. In fact, all positions on our spectrum (except possibly a tiny fringe at the right edge) say, "Not now." But while Group 1 says, "No germline intervention, on principle," Group 2 says, "No germline intervention now, but possibly or possibly not some day."

The World Council of Churches advocates a ban on germline intervention "at the present time," and "encourages the ethical reflection necessary for developing future guidelines in this area" (World Council of Churches 1989, 2). The National Council of Churches urges "extreme caution," then continues: "Genetherapy of germ-line or sex, cells of human embryos—if ever practicable—will deserve stringent control" (National Council of Churches 1986, 4, 13). That leaves "the window open a crack."

Several Protestant denominations in North America have issued policy statements or study documents: The United Methodist Church, The United Church of Christ, The Episcopal Church, The Presbyterian Church, three cooperating Lutheran denominations, and The United Church of Canada. All those churches point out the deep ethical problems involved in the new genetics, but none—except *The Methodist Discipline*, as already noted—make a total rejection of germline therapy. All belong in Group 2 of our spectrum.

The Catechism of the Catholic Church repeats earlier papal teaching: "Certain attempts to *influence chromosome or genetic inheritance* are not therapeutic but are aimed at producing human beings selected according to sex or other predetermined qualities. Such manipulations are contrary to the personal dignity of the human being and his integrity and identity . . ." (*Catechism* 1994, paragraph 2275). That statement carefully avoids rejecting germline interventions that are clearly therapeutic. It may be that other Catholic teachings on the immorality of artificial insemination and test-tube babies (*Catechism* 1994, paragraph 2377) will, in effect, mean rejection of all possible methods of germline therapy. If so, the Catholic position, for all practical purposes, falls into

A Crucial Issue: Modification of Germline Cells

Group 1. But, unlike Group 1, there is here no *categorical* rejection of germline interventions that are therapeutic.

3. Restrained expectation.

Group 3 on the spectrum expects some genuinely beneficial results from germline interventions. The late Bernard Davis of the Harvard Medical School represented this position. He advocated continued research with the hope, although not the assurance, that parents could have children "without condemning them to inherit a particular defective gene." Using diabetes as an example, he found the possibility of germline therapy to be beneficial, provided "harmful side effects" can be avoided. It would be superior to somatic therapy that could help an individual, who would still pass on the deficiency to offspring.

The expectations of Davis were restrained in two respects. First, he was not predicting actual success in germline therapy; he considered the prospects valuable enough to justify continuing research. Second, he thought it "extremely unlikely that we will also be able to improve behavioral traits in this way, for the polygenic nature of these traits suggests that they would require extensive replacements of DNA for any substantial and predictable changes" (Davis 1973, 21–22). But he did not rule out even this possibility, and he strongly endorsed continuing research.

A sampling of the extensive literature on this subject indicates that the dominant opinion among geneticists gravitates toward this position. Although I have quoted statements of Davis from 1969, the later discussions, taking account of more recent developments, see the basic issues pretty much as Davis saw them. An example, representing a considerable international consensus, is the report of UNESCO's International Bioethics Committee, proposing a ban of germline intervention for purposes of "enhancement," but endorsing it (with due cautions) for therapeutic purposes: "If it is morally acceptable to cure a condition by somatic gene therapy, then why not to eliminate it by the germline?" (*Nature* 1994, 369).

4. Enthusiastic anticipation.

Group 4 represents the most fervent hopes of genetic alteration of the human race. The Jesuit priest and paleontologist Pierre Teilhard de Chardin expressed this opinion in rhapsodic terms. In 1946 he foresaw "control of heredity and sex by the manipulation of genes and chromosomes," and in 1948 he looked forward to eugenics "as a controlled process in the proportion most beneficial to humanity as a whole" (Teilhard 1964, 144, 234).

Teilhard put his hopes in the context of an invincibly optimistic metaphysics in which he saw even the atomic bomb as a prognosticator of a better world. For a time, he won wide popular acclaim. Teilhardians talked exuberantly of a perfection of human nature, of the humanization and hominization of the universe, of humanity taking charge of its own evolution.

Today that language sounds a little wistful and quaint. But the popular culture craves such assurances. Marlon Brando, agonizing over the world's cruelties, within his family and the larger society, finds some solace in the confidence that genetic changes will some day overcome human "self-destruction" and "the will to kill" (Brando 1994, 464–65).

Biologists are, for the most part, far more sober. Robert L. Sinsheimer, at least for a time, foresaw the "eugenic transformation of our species," then immediately recognized the problem: "The horizons of the new eugenics are in principle boundless. . . . But of course the ethical dilemma remains. What are the best qualities, and who shall choose?" (Sinsheimer 1987, 137, 145).

Edward O. Wilson, who made famous the word sociobiology, says: "The human species can change its own nature." It may

A Crucial Issue: Modification of Germline Cells

"press on toward still higher intelligence and creativity." "New patterns of sociality could be installed in bits and pieces." "But," he continues, "we are talking here about the very essence of humanity." Wilson is not at all sure that such changes are possible or that people will choose to make them. But he expresses a commitment to hope (Wilson 1978, 208–209).

Robert Shapiro, in the widely-read *The Human Blueprint*, concluded: "By applying our consciousness to evolution, we should be able to make our position and that of life, in general, not only more secure, but ultimately much better." Taking account of varying human preferences and values, he said: ". . . the human race may separate eventually into a number of subspecies (perhaps separated by national borders) each following its own vision as to the best biological future for our species" (Shapiro 1991, 374). Some of his critics saw more danger than he in this proposal. Some noticed a troubling resemblance to the "clannishness" advocated by Murray and Herrnstein a few years after Shapiro's suggestion. Others wondered whether, if Iraq or Iran (for example) should choose development of a more militant subspecies, other countries would placidly approve.

I have been reporting a few samples from a wide spectrum of opinions. It is now time to dig into them and look for the fundamental scientific and ethical issues implicit in the opinions. Following the method described in Chapter 4, I distinguish between scientific judgments and judgments of value and purpose. I recognize a qualified autonomy in each realm, at least to the extent that neither can prescribe conclusions to the other. But I recognize also a constant interaction, evident in most opinions and beliefs about appropriate and valid policies.

Scientific Issues in Current Debates

Some issues in the current controversies are scientific. Science alone will not determine what a society ought to do, but all issues of policy have a scientific component. In arguing for a qualified autonomy of science, I have affirmed that, within a defined sphere, no outside force—political, ethical, or religious—can determine scientific truth. For an example, I turn again to the long and dramatic search for the gene for Huntington's disease. Many forces motivated the search: the joy of scientific discovery, but also the desire to find ways of avoiding or healing a dreaded disease, the availability of funding for that particular project, the ambition of researchers. There would have been no discovery of the gene without the value-driven efforts and commitments of many people. Furthermore, the values and concerns of individuals and social groups will guide the uses of the new knowledge. But the news that the gene is located on chromosome 4 was a scientific announcement. No extra-scientific force, no hopes or fears can change that. If the discovery is modified by future research, it will be scientific research, not an act of Congress or artists or religious authorities.

On the issues of germline modification, there are some questions that are scientific. I can here identify three, with the aim of showing both the autonomy of science and the way in which that autonomy is qualified when we move from scientific determination to questions of policy.

1. Is germline modification feasible?

Here we start with the fact that experiments have practiced germline modification on mice and rats. That is evidence that, in some sense, the process is possible. In the early 1980s a rat growth hormone was injected into fertilized mouse eggs; some of those eggs developed into mice double normal size (Shapiro 1991, 367). That does not show that anybody could, or would want to, inject elephant DNA into fertilized human eggs and produce 500-pound

A Crucial Issue: Modification of Germline Cells

linebackers for the National Football League. It shows only that some kinds of germline intervention are possible.

More recently, experiments at the University of Pennsylvania have altered the sperm of mice and have applied for a patent—another controversial issue—on the process. They foresee new possibilities in the breeding of animals, and they point out that any human application raises ethical issues. Given the many genetic similarities between mice and humans, there is little doubt that human germline intervention of some sort is possible. But before pronouncing it feasible, we must investigate the risks.

2. What are the risks?

The fertilized egg has been described as a millionth the size of a pinhead. To intervene with enough precision to insert a chosen gene at the right point, then to expect that the egg would accept and integrate the new gene, is a bold hope—maybe an invitation to error. According to C. Thomas Caskey, "Current gene transfer and replacement technologies have low success rates (between one in a thousand and one in a hundred thousand) and high rates (between one in ten thousand and one in a million) of illegitimate recombinations, in which the gene inserts itself into the wrong place, sometimes in the middle of another gene" (Caskey 1992, 129). Germline treatment of mice has multiplied the rate of cancer in the new generations, and injection of a cancer-preventing gene into embryonic mice has led to many cases of blindness and other deformations.

There is the further risk that germline intervention could introduce a recessive liability, which could remain undetected for several generations. Then some descendant of the original subject would mate with somebody else with the same recessive trait, and the offspring would show a disease, perhaps hitherto unknown. By that time, the recessive trait might have spread to many other people.

To consider another risk, recall (Chapter 2) that a single gene may influence twenty characteristics of a person. If we "correct" or replace one defective gene, we need to ask what other unwanted changes we may cause.

We might ask the hopeful question: If errors introduce liabilities into the lives of mice or people, might genetic science correct those errors by new treatments, somatic or germline? Perhaps, perhaps not. It is easier to cut off a person's head than to restore it. Genetic errors, likewise, may or may not be correctible.

No life is risk-free, and people constantly accept some risks. What science can do is give estimates of risks and benefits in any procedure. Whether people—individuals, medical institutions, and governments—accept or permit specific risks is not a solely scientific question. Risks must be evaluated as well as measured. I shall come to that issue soon.

3. Does germline therapy do anything for the patient that somatic therapy cannot do?

We frequently read that genetic treatment of somatic cells can do everything that treatment of germline cells can do *for the patient*. But another opinion is that in a very few cases germline treatment might prevent a disease not subject to somatic treatment. That argument may, sooner or later, be settled on scientific grounds. Whichever way it turns out, the somatic treatment is less risky. Rarely is the germline treatment likely to be better for the patient.

But the advantage of germline treatment, if it becomes feasible for humans, is that it eliminates the ailment from the continuing germline. If I, for example, have a genetic ailment that can be treated somatically, I am the only one to benefit from the

A Crucial Issue: Modification of Germline Cells

treatment. I can still pass on the liability, as a dominant or recessive trait, to my children. They or their heirs *ad infinitum* might need the same somatic treatment that helped me. But if, way back at my conception, I had received successful germline therapy, my vulnerability to that specific disease would be removed from my descendants in perpetuity. Occasionally we hear speculations about "cleansing" the human race, by germline therapy, of some hereditary ailments (diabetes, cystic fibrosis, or Tay-Sachs). But scientists estimate that such a program, even if it should prove feasible, would involve mass compulsory treatments over many centuries.

Geneticist Bentley Glass asked the question, "How can we predict the value of a gene in environments other than those in which it is now known to exist?" (Glass 1965, 63). Soon after, Dobzhansky asked, "Does anybody know what will be best for mankind centuries or millennia hence?" (Dobzhansky 1967, 411).

The genes that cause ailments sometimes also have survival value. The frequent example is sickle-sell anemia, where the recessive trait offers some protection from malaria, accounting for the persistence of the trait in malaria-prone regions. A double dose of the gene, from both parents, produces the serious illness. Such examples lead to warnings against trying to "purify" human heredity by eliminating one peril after another, with a consequent "narrowing of the gene pool," perhaps to the disadvantage of generations yet to come. It is the diversity of the gene pool that helps the human race survive and adapt to a great variety of situations.

E.O. Wilson writes, "No artificially selected genetic strain has, to my knowledge, ever outcompeted wild variants of the same species in the natural environment" (Wilson 1989, 114). Gardeners, who pamper their hybrid tea roses and watch wild roses grow rampant, have often guessed this. A human race, enhanced by genetic modification, might face a more precarious existence on a planet that, over the centuries, can expect many changes in climate and environment.

Even so, most parents would prefer to spare a child a serious ailment rather than perpetuate a disabling gene that might, or might not, in some unforeseen future have a survival value for descendants in a transformed environment. *If* germline therapy can eradicate Huntington's disease or Tay-Sachs or several others, and *if* it can do so without harmful consequences—two monumental *ifs*—there are strong reasons for keeping the door open to germline therapy.

In looking at these three questions, we see again that in each case there are some questions that can be settled only on scientific grounds, if at all. Our hopes, our compassion, our ethical sensitivities cannot prejudge or nullify the findings of science. And science makes an utterly necessary contribution to responsible action. But in each case the scientific data alone do not dictate a policy. Those data enter into ethical judgments, but there they interact with human sensitivities, values, and concerns that are extra-scientific.

All of us who make personal decisions or try to influence public policy must, at some point, rely on the best scientific judgments we can get. Our hopes and fears do not change the biochemical processes of DNA. But even here, we are not simply passive recipients of information. We can address to professional scientists questions that may not be foremost in their minds. Also, to return to a theme of Chapter 3, we must ask how far the values and ideologies of experimenters influence their judgments.

The published studies of risks are an interesting case in point. The gung ho adventurers among scientists tend to minimize risks. The more restrained highlight the risks. Both groups, insofar as they are genuinely scientific, are willing to investigate evidence that may modify their judgments. But when various estimates are available, as is usually the case in exploratory science, personal

A Crucial Issue: Modification of Germline Cells

temperament and ideology enter into judgments, often decisively. It is not that all participants perceive the same evidence, then evaluate it differently. Rather, our differing sensitivities and ideological persuasions often determine what evidences seize our attention.

That is why I said earlier, and now repeat: The second most important understanding in the relation between science and values is to see the distinction between the two, refusing any imperialism of either. The most important understanding is to see their continuous relationships and interactions.

Issues of Purpose and Values

In focusing attention on scientific questions, we found that they led constantly to issues of purpose and values. Shifting our focus to questions of value, we can never ignore the scientific questions, but we have to incorporate them in a wider context.

1. Risk, revisited

The calculation of risks is a matter of examining evidence and combining it with arithmetic. The projection of risks is a process of estimates, and perfect accuracy is not possible. But the evidence is available to many observers, and the arithmetic is either right or wrong. So a scientific estimate of probabilities is possible, although wariness of ideological skewing is always important.

Scientific probability is the basis of the cost-benefit analysis that all of us use from time to time. Car drivers frequently choose between different routes: one is shorter, but has more traffic jams; or one is better on snowy days, the other on dry days; or one is faster, but the other is more scenic. Drivers weigh the costs and benefits of each route and make their choices.

Whole systems of ethics have been built on cost-benefit methods. Jeremy Bentham's utilitarianism tried to do that. With a quasi-sophisticated way of quantifying pleasure and pain, Bentham urged the maximization of pleasure and the minimizing of pain (Bentham 1789). Then his successor, John Stuart Mill, while still claiming to be a utilitarian, pointed out the flaw in the system. "It is better to be a human being dissatisfied than a pig satisfied; better to be a Socrates dissatisfied than a fool satisfied" (Mill 1861, Chapter 2). The peculiar nature of selfhood transforms our reasoning. We face what we call, in differing languages, the worth of the person, the dignity of the individual, the inviolability of selfhood, the image of God in human life. We do not undertake portentous medical experiments, least of all genetic experiments, with the hope that we'll distort some lives but will help a greater number.

An ancient adage of medical ethics is, "Do no harm." It can be challenged. Is it justifiable to risk doing harm for the sake of a greater good? The seriously sick patient may want an experimental treatment that offers some hope of healing, even though it is risky. Thus people with AIDS sometimes ask for drugs that have not yet been tested and approved by the Food and Drug Administration. The standard rubric for risky medical care is the informed consent of the patient. Sometimes people volunteer for treatments, even though there is no personal advantage, because they want to contribute to finding a cure for others. That is regarded as noble conduct. But it is not noble to impose risks on people, without their consent, for the sake of medical progress.

But informed consent cannot be an absolute requirement. Infants and patients in a coma cannot give informed consent. Our society is struggling to find the right protections for such people, sometimes by naming a surrogate to give or withhold consent. The living will is a device by which individuals designate others to act in their stead when they are incapable of informed decision. But no fertilized egg ever executed a living will.

In the case of germline treatment, informed consent is

A Crucial Issue: Modification of Germline Cells

totally impossible. The future generations, who may benefit and are surely put at risk, cannot be consulted. That counts seriously against it. We cannot ethically accept the prospect of harming or destroying many lives along the way to achieve greater gains for generations to come. Yet the day may come when we cannot ethically leave in jeopardy those future generations by our failure to act on their behalf. Our opportunity may be linked to a tragic dilemma—a sign of the tragic element in all serious ethics. That is one reason why ethics, in its depth, always fades into some larger vision of life and death, usually called religious.

But, to pause a moment longer on the issue of informed consent, this requirement cannot be absolute. None of us gave informed consent to our own birth. None of us gave informed consent to our individual genomes. Yet most of us are glad that we were born. And whatever discontents we have with our genetic inheritance, most of us are moderately, often excessively, pleased with ourselves.

Risk analysis in an ethical context leads to the conclusion: any future experimentation on germline therapy faces an immense burden of responsibility. That burden does not, of itself, constitute a prohibition for all time.

2. The issues of eugenics

Germline intervention is future-oriented. Whereas somatic therapy aims to help a single individual, germline therapy aims to improve the inheritance of the persons who will descend from the present individual. That aim gives it both its persuasiveness and its intimidating power.

One study document, pointing to its possible benefits, said: "By overcoming a deleterious gene in future beings, the beneficial effect of such changes may actually be magnified" (World

Council of Churches 1981, 6). The same document warned against the perils in imposing biased norms on generations to come.

Germline intervention is a eugenic practice, in the strict sense of that term: the attempt to improve human heredity. And eugenics, in our cultural climate, has become almost a nasty word. There are reasons for this. The term originated with Charles Galton, a cousin of Darwin. A eugenics movement developed in England, Germany, and the United States. In the United States, to pick up a theme from Chapter 5, it was often a favorite cause of middle-class intellectual people who saw their own kind as desirable models for the world. They often wrapped into one gorgeous package their own "liberal" idealism, genetic foolishness, and prejudice against foreign immigrants and people of "inferior" physical health and mental ability. Some of the courageous pioneers of feminism and other democratic causes fell into the trap. That gives force to the warning of Marshall Nirenberg, who won the Nobel prize (for "breaking the genetic code," as some have put it): "When man becomes capable of instructing his own cells, he must refrain from doing so until he has sufficient wisdom to use this knowledge to the benefit of mankind" (Nirenberg 1969, 633).

In Germany the Nazis took up eugenics with their perverse exaltation of "Nordic" types, their vicious destruction of the Jews, and their cruelties in medical experimentation. They gave eugenics so bad a name that, in the recent past, to label any practice *eugenic* has usually been to condemn it. The "liberal," democratic ethos prescribed an emphasis on the possibilities of education and culture to improve society, and a consequent minimizing of heredity. But, with the discovery of DNA and consequent developments, a new attention to heredity crept back into the public discussion, as we have seen. Some proposals were eugenic in nature, although with far greater scientific understanding than in

A Crucial Issue: Modification of Germline Cells

the past and with an almost ritualistic avoidance of the term *eugenics*.

Social location has much to do with the beliefs of people about the uses of genetics. In four international conferences, I have found Europeans—especially, but not only Germans—to be notably more worried about the misuses of genetics than Americans. The dread Nazi legacy hovered over conversations. I have also found the Third World (or the Two-Thirds World) more worried than the North Atlantic World. Asians, Africans, and Latin Americans are likely to suspect that genetics is one more plot of the wealthy, high-tech nations to dominate the world. It is easy to find exceptions to this generalization; Americans frequently sound alarms. But for the most part we Americans are more confident of our ability to handle new powers wisely than are others. Their ideology, supported by their memories and observations, is more suspicious than ours.

Today the one major nation that openly espouses eugenic practices is, curiously, China. A Maternal and Infant Health Care Law bans marriage of people with specified genetic diseases, unless they agree to sterilization or long-term contraception. The system also puts pressure on parents to abort a fetus with an abnormality. It outlaws sex screening and the consequent abortion of female fetuses, as I mentioned in Chapter 2; but that provision is not very effectively enforced. Repeating some mistakes of the West, it attributes to heredity some ailments that are due to poverty and inadequate diet. The present policies are a total reversal of China's former Marxist ideology, which denigrated heredity and counted on social transformation to solve ills of overpopulation and health.

The United States, with its greater affluence and its glorification of individualism, is not greatly tempted by the governmental policies of China. Our comparable prejudices are more likely to express themselves in hostility to immigrants and the poor, often with a barely-camouflaged eugenic bias.

Prejudices, latent or overt in every society, if expressed in eugenic planning, lead to fearful forebodings. Here I find it useful to distinguish between "hard" and "soft" eugenics. The *hard* style acts with blatant cruelty. James Watson, almost the high priest of American progress in genetic science, sounds the warning: "We have only to look at how the Nazis used leading members of the German human genetics and psychiatry communities to justify their genocide programs, first against the mentally ill and then the Jews and Gypsies. We need no more vivid reminders that science in the wrong hands can do incalculable harm" (Watson 1990, 44–49). Herman Kahn, although famous for his technological optimism, did a double take when it came to genetics. In an interview with the press, he said: "Genetic engineering has in it the makings of a totalitarianism the like of which the world has never seen" (Kahn 1971, 24).

Hard eugenics is centralized and compulsory. It could tempt dictators to aim at the production of super-warriors. Will nations or factions that now control great power use new techniques to increase their power? We might already wish that humankind had never discovered the techniques of chemical and biological warfare, which have lately been adapted to the destruction of innocent civilians by angry saboteurs.

Soft eugenics, by contrast, means a quiet enforcement, perhaps even by persuasion, of "normality" on heterogeneous populations. Salvador Luria, the MIT biologist, says: "The real danger today is the possible emergence of an establishment program to invade the rights and privacy of individuals, whether in the area of sexual preference, or right to abortion, or drug addiction, under cover of beneficent eugenic intervention." He is less worried by a Nazi-style program than by "a kinder, gentler program to 'perfect' human indi-

A Crucial Issue: Modification of Germline Cells

viduals by 'correcting' their genomes in conformity, perhaps, to an ideal, white, Judeo-Christian, economically successful genotype" (Luria 1989). Ruth Hubbard and Elijah Wald express comparable concerns (Hubbard and Wald 1993). Occasional suggestions that a genetic analysis may some day become as common as a vaccination (Asimov 1962, 181) may comfort some people and bring a shudder to others.

C.S. Lewis pointed out that eugenics means that the present generation, or actually a minority within it, shapes the destiny of an indefinite future—a few hundred people thereby "ruling over billions upon billions" yet to come. It was in this context that he made the statement, so often quoted: "Each new power won by man is a power *over* man as well" (Lewis 1947, 71).

But such generalizations need examination. One of the great biological achievements of the present generation has been the elimination of smallpox—probably for all time to come, although nobody can be quite certain of that. It is hard to imagine that future generations will resent this power that our generation has exercised over them.

3. More modest expectations

If we can separate germline modification from the larger eugenic ideology—as I think we can, though with great difficulty—questions remain. One comes from the inflated expectations evident in the popular press. There is talk of babies made to order with "designer genes," to use the inevitable phrase. There is more talk of a population with heightened physical capacities and intelligence. Here we need the warning of geneticist Theodosius Dobzhansky: "Human cultural capacities can hardly be generated by so crude a mechanism as one-to-one relationships between a gene and a trait or capacity. . . . There is no such thing as a gene for self-awareness, or for consciousness, or for ego, or for mind. These basic capacities derive from the whole of the human genetic endowment, not from some kind of special genes" (Dobzhansky 1967b, 72–73).

Genetic power, like all power, calls for constant vigilance. So far as the healing of destructive diseases is concerned, the vigilance has to do with risks of unintended harm and with issues of just access to new, probably costly developments. The more the new powers move from *healing* to *enhancement* to *distortion*—a theme of Chapter 5—the greater is the call for vigilance.

We have been examining the interaction of scientific and evaluational judgments in assessments of policies for germline interventions. Every scientific proposal raises questions of values; every valuation incorporates scientific data and concepts. An interesting example of scientific-ethical collaboration is the work of W. French Anderson and John C. Fletcher. Anderson has been the ambitious "point man" in genetic therapy. Fletcher is a theologian and ethicist. The two have worked together in looking at the feasibility of germline therapy. They assess the possibilities and risks of proposals and conclude that we are not ready to proceed until extensive experiments on animals have established the feasibility and risk of new procedures—a scientific issue. But even if that takes place, they see extensive social and ethical issues that reach far beyond clinics and laboratories. Hence they call for wide public education and discussion, with no action until "a solid societal consensus" emerges. "This therapy," they say, "will affect future generation. . . . The gene pool is a joint possession of all members of society. Clinical trials should not begin until the public is aware of the issues and has had an adequate opportunity to express ethical and social views in a national [perhaps they would agree, international] forum" (Fletcher and Anderson 1992, 29).

If all that takes place, they are ready to endorse germline therapy to remedy or prevent "disorders with the greatest magnitude of suffering, pain, and early death." They oppose attempts "to

A Crucial Issue: Modification of Germline Cells

enhance particular traits that have little or nothing to do with disease" (Fletcher and Anderson 1992, 31).

Those who endorse and those who differ with specific opinions of Fletcher and Anderson would be wise to heed their double emphasis on scientifically-competent procedures and ethical responsibility to a wide public.

Playing God

The warning against playing God pops up frequently in conversations about genetics. It carries diverse meanings. For some, it has a strong theological meaning. It recalls the tower of Babel from the Hebrew Bible, the story of Prometheus from Greek mythology. For others, it implies a transgression into the ways of nature that have produced this immensely complicated human genome, so intimately related and yet so different from the genomes of bacteria and animals. For still others, it is only a metaphor for rash and dangerous acts, staking a valuable genetic inheritance on experiments that might go wrong. For almost all, it has the connotation of touching on the mysteries of human selfhood, of living nature, even of the cosmos and the ultimate power that created it or operates within it.

Even in so secularized a society as ours, the phrase has a dramatic resonance. It echoes in the titles of books like *Who Should Play God?* (Howard and Rifkin 1977). It leaps out of the report, *Splicing Life*, of the President's Commission for the Study of Ethical Problems in Medicine, which devotes several pages to "Concerns about 'Playing God.'" Granted, the commission was, at the request of President Carter, responding specifically to an inquiry from Jewish, Catholic, and Protestant religious associations. (See Chapter 4.) Even so, it is remarkable, in a society marked by separation of church and state, that a governmental commission, composed mostly of scientist and physicians with one Rabbi, should give its opinion on a theological issue, documented with a footnote to Pope John Paul II (*President's Commission* 1992, 56).

Although the question about playing God should give us pause, it does not answer itself. Theology, even in all its diversity, has traditionally taught that God acts both through the processes of nature, which long antedated humankind, and through human beings, who sometimes serve and sometimes resist the divine will. It has also taught that God is the source of healing and that human healers serve God. That basic conviction does not of itself tell whether specific biomedical acts are a sign of God's activity or acts of defiance. To make that judgment requires the kind of reasoning that has governed most of this chapter.

Yet there is one more general consideration in this area. I touched on it in Chapter 5, when I discussed manipulation, with an eloquent quotation from Margaret Mead. At what point does a human managerial mentality change our perception of human life? A traditional hymn, based on Psalm 100, says of God: "Without our aid he did us make." Certainly the galaxies, the earth, and many millions of species including the human species were made without our aid. But for centuries, the nurture and healing of human life has been done with our aid. Are we approaching a time when we may regard the newborn baby less as a gift than as an artifact of our production? Factually, I see no evidence that genetics is about to turn into a combination of architecture and carpentry. However, attitudes outrun facts, and a take-charge attitude, displacing all reverence before the mysteries of and behind nature, would greatly change our perception of ourselves and our world—not, I think, for the better.

However, the immediate question is germline therapy. I am more interested in defining the issues, so that many people may think them through, than in delivering verdicts. However, I need not be so cowardly as to conceal any judgments of my own. My own

A Crucial Issue: Modification of Germline Cells

opinion favors, for the immediate future, genetic healing of somatic ills but not germline therapy. I am impressed by the judgment of Robert Cook-Deegan that the leap from the animal experimentation now underway to human applications is very great, because of the "highly distinctive" nature of "human brain development" (Cook-Deegan 1994, 218). I take seriously the judgment of John Maddux, the distinguished editor of *Nature*: "The side effects and the uncertainties [of direct germline manipulation] will always be greater than those of the techniques already available" (Maddux 1994, 15). But I stumble over his word always. With Cook-Deegan, I want to "keep the window open a crack." I think the World Council of Churches and the National Council of Churches were right to issue grave warnings, but to stop short of an ethical prohibition for an unforeseeable future.

Chapter Seven

Coda

The pathway of this investigation has been fairly simple. Not entirely so. It has led through explorations into intricate discoveries about the nuclei of human cells and the transmission of hereditary traits. It has required a few digressions into byways, like the meaning of intelligence and its relation to heredity. But the main path, I hope, has been clear.

My initial concern has been the relation of science and ethics in policy decisions, including decisions of persons, families, and social organizations, but with an emphasis on public policy. I have maintained that every such decision depends on (1) *science*, understood as a body of verifiable knowledge and as a set of methods and concepts, and (2) those human purposes, commitments, and goals that I have loosely called *values*. Every policy involves both. It also involves (3) *politics*, which in its exercise of power both exemplifies and distorts scientific understanding and values.

I have argued for a qualified autonomy of both science and values. The autonomy means an insistence that neither can dictate conclusions to the other. Our commitments do not themselves inform us that human life has emerged after two or three or four billion years of biological evolution on earth, that the human genome includes forty-six chromosomes, or that the gene for Huntington's disease is located on the tip of chromosome 4. Science has made these discoveries. They appear to be well verified. If they should be modified, as scientific knowledge often is modified, that will be the result of more scientific research, not of dictation from politicians, corporations, philosophers, poets, or theologians.

But, equally important, science cannot prescribe our ultimate commitments. It can show us ways to accomplish some goals. It can open up new possibilities and thereby help us redefine and expand our goals. But it cannot, of itself, tell us the value of human life, the meaning of human dignity, or the worth of courage and altruism. It can make possible the destruction of civilization in nuclear war or, perhaps, the alteration of human nature by genetic means. But the possible is not always the desirable. Every projected achievement requires evaluation.

Hence I have used the word autonomy for both science and ethics, while insisting that the autonomy is qualified. Although neither can dictate conclusions to the other, the two constantly interact.

So I have argued that science, as a human enterprise, is value-infused. Some values are inherent in science, especially integrity, accuracy, and openness to evidence. Beyond that, science — above all, big science, which is costly science — expresses the values of governments, corporations, foundations, and universities that sponsor it. The choice of priorities for research, the persistence in the face of obstacles, the zeal to prevent or heal age-old diseases — these are values that guide research. Only in one narrow, though important, respect is science value-free: investigators must, if they are competent scientists, follow the evidence where it leads, even though the evidence may refute cherished hypotheses and disappoint ardent expectations. Therefore I oppose two common opinions: one holds that science as a total enterprise is value-free; the other sees science as solely an instrument of power and denies that verification, at some utterly important points, is a value-free process (except for the inherently scientific values of accuracy and honesty).

Similarly, I have argued that the autonomy of values is qualified. No set of coherent values descends from heaven or some realm of pure intellect, simply to be applied to the confusions of human life. Human problems involve a variety of values, sometimes in tension or conflict: justice and mercy, economic productivity and ecological concerns, short-range and long-range benefits, self-regard and concern for others, individual freedom and the demands of society. Common experience, sometimes helped by science, helps us sort out the values, sometimes subordinates some to others, sometimes discovers new possibilities of transcending old conflicts. So the autonomy of values is qualified. But human values usually rest on some fundamental insights and convictions—e.g., about human dignity, human rights, freedom-in-community—that are not likely to be shaken by the latest news from a laboratory.

Both science and values, I have further maintained, are rooted in faith. Any formulation of faith rises out of experience and is subject to reformulation in the light of future experience. But experience itself is appropriated and guided by faith. Faith, of one kind or another, is a presupposition of all talk of science or values. Science rests on what George Santayana called "animal faith," the almost unthinking faith that our senses and our minds, for all their fallibility, lead us to reliable knowledge. It continues with that faith, described by William James, in the supreme good of the search for knowledge and the correction of error (see Chapter 4). Just as obviously, all our values are expressions of a faith, usually that life is worthwhile, that courage is a virtue, that suffering is something both to be overcome and to be endured, that justice requires each of us to relate our claims to the claims of others, that persons are incomplete without loyalties that transcend individual interests. I have argued that the specific historic faiths are to be treasured, but that public policy, in a pluralistic society, must appeal to widely-shared commitments that underlie and support political processes.

Thus I have kept in mind, throughout this investigation, the three terms of my subtitle: *science, faith*, and *politics*. A fourth term, not in the title, has demanded attention: *ideology*. It is the picture of the world that, often surreptitiously, guides action. It is a composite of science, ethics, and faith—and a lot more: attitudes, interests, common sense and common nonsense, passions and prejudices. It is also the lens through which we see reality, the gestalt within which we incorporate new perceptions and insights. It always works from a point of view; that is its opportunity and its problem. Without a point of view, no one can recognize anything. But every point of view is particular and excludes other points of view. So ideology is a peril as well as a necessity. Every participant in public policy had better recognize ideology. Rather than claim to rise above ideology—a self-deceptive pretense—each of us had better take account of diverse ideologies. Policy formation is a continuous quest, with moments of perplexity, achievement, new perplexities, and revision of past achievements.

The Human Genome Project has been the case study for this inquiry. It is a significant case study, both because of its urgent importance and because it involves all the theoretical issues that I have just been recalling. It presents new opportunities and perils to humankind. It calls for an exploratory ethic, but that need not be a capricious ethic or an ethic of mere expedience. It involves new powers, never before known to humanity. It also warns us that not everything that can be done ought to be done. It requires us to relate pioneering science to ethics and to the faiths that infuse our world, with constant awareness of political processes and of ideology. It requires the relating of the most fundamental traditions of faith and ethics to some quite nontraditional possibilities.

For the sake of a reasonably sharp focus, I have narrowed my investigation to the issues of genetic therapy. I have omitted, except for passing mention, some important issues:
- What shall we do with the troubling economic issues? Will costly new procedures be available only to the affluent? Will health insurance agencies, whether for-profit or governmental, decide who gets access? What about questions of patents, which may be extremely profitable and may limit the benefits of new procedures?
- What about the privacy of genetic information? Will insurers and employers, who now often require medical examinations, also require genetic information? Might an airline require genetic tests of a pilot that it does not require of baggage-handlers (as it already requires more rigorous somatic tests of pilots)? Will the public, already interested in the health of political candidates, demand genetic information about future presidents? Will prospective parents have a moral, maybe even legal, obligation to disclose to each other genetic liabilities that endanger offspring? Might governments require genetic testing as a condition for issuing marriage licenses, as they already require tests for syphilis? Will laws establish a new category of the genetically disabled, with both rights and privileges, along with liabilities, comparable to present categories of the physically disabled? Will a new caste system, based on genetic liabilities, arise?
- What effects will genetic knowledge have on criminal justice? DNA "fingerprinting" has led to conviction in court of some

suspects and to exoneration of others. Will the plea of "not guilty by reason of insanity" be expanded to include, "not guilty by reason of genetic determinism"?

More issues will arise, as important as the ones I have chosen for attention. My neglect of them throughout most of this book does not mean that I think them unimportant. On the issue of genetic therapy, which I have chosen to examine, I have looked at some of the questions, both immediate and on the horizon, of genetic modification. I have assumed that genetic therapy of somatic cells, already in experimental stages with humans, can be judged by the same criteria already used in all experimental medicine. Hence I have focused attention on the highly controversial issue of germline therapy. Here my main concern has been not to argue for my particular judgments, but to encourage wider discussion of the issues in the context of awareness of the relations of science, values, and politics. But, rather than be coy, I have stated my present opinions. In agreement with almost universal informed opinion, I reject germline modification at the present time. In opposition to some opinion, I want to keep the window open for research and possible action, if the time comes when such intervention may be possible for relief of serious ailments. In opposition to some other opinions, I doubt that attempts at enhancement, or even more ambitious attempts at redirecting evolution, will become possible (a scientific issue) and I deeply doubt that it will be ethically responsible (an issue of values). But nobody now is capable of foreseeing what may become possible or issuing ethical edicts for all time to come.

Repeatedly in this investigation I have come to the edge of far larger questions. The new genetics inevitably draws attention to old, old questions about the place of human beings in the whole of things. It tells us of our close kinship with chimpanzees and mice, yet heightens our power to master some of the powers of nature that have constrained us. It increases our realization that natural processes, indifferent to us, shape our lives; yet it gives us new freedoms to influence and sometimes direct those processes. It accentuates our awareness of both our power and our weakness.

Blaise Pascal (1623–62), the one person in all history who won great distinction as both scientist and theologian, did his scientific work in physics and mathematics, with no knowledge of genetics. Yet one of his persistent themes gains new persuasiveness in this time of genetic prowess. No one saw more forcefully the grandeur and the weakness of the human person:

> Man is but a reed, the most feeble thing in nature; but he is a thinking reed. The entire universe need not arm itself to crush him. A vapour, a drop of water suffices to kill him. But, if the universe were to crush him, man would still be more noble than that which killed him, because he knows that he dies and the advantage which the universe has over him; the universe knows nothing of this.
>
> All our dignity consists, then, in thought (Pascal 1670, Fragment 97).

But Pascal, almost like a post-modernist today, knew also the weakness of thought: "We burn with desire to find solid ground and an ultimate sure foundation whereon to build a tower reaching to the Infinite. But our whole groundwork cracks, and the earth opens to abysses. Let us therefore not look for certainty and stability" (Fragment 72). Again, he wrote:

> All the dignity of man consists in thought. Thought is therefore by its nature a wonderful and incomparable thing. It must have strange defects to be contemptible. But it has such, so that nothing is more ridiculous. How great it is in its nature! How vile it is in its defects! (Fragment 365).

About a century later Alexander Pope (1688–1744) discovered Pascal with fascination. Less profoundly but with skill in rhyme and rhythm, he put the same theme in verse in "An Essay on Man" (Pope 1734):

> Created half to rise and half to fall;
> Great lord of all things, yet a prey to all;
> Sole judge of truth, in endless error hurled;
> The glory, jest, and riddle of the world!

Today those paradoxes are all heightened. We have won powers incredible in any past. Yet we know better than any past that "the entire universe need not arm itself to crush" us; a gene, a nucleotide or two can do it. We can welcome and exult in new power, especially the expanded abilities to prevent and heal destructive illnesses. But more poignantly than ever before, we know our frailty. We are vulnerable beings, destined for death. We had better learn to cherish:

> The still sad music of humanity,
> Nor harsh nor grating, though of ample power
> To chasten and subdue.
> (Wordsworth 1798, lines 91–93).

Like it or not, we must meet sadness with whatever courage and faith are given us. That "still sad music" can crash into a cacophony of despair; it can, miraculously, soar in an "Ode to Joy" or a "Hallelujah Chorus."

We have earlier seen how frequently the phrase, "playing God," enters into the public discussions of genetics, even getting a place in a government-sponsored publication. In biblical language, the human creatures are placed in a garden, granted a dominion over nature and a responsibility to care for the garden. The same human powers later lead to the construction of a tower of Babel, ending in retribution for pretentious pride and the confusion of tongues. Only a rare union of courage, modesty, and wisdom can

CODA

begin to discern the differences, whether stark or subtle, between creativity and destructiveness.

As I conclude this inquiry, the Toyota corporation is running advertisements in several magazines, with the aim of showing that Toyota benefits the American (not just the Japanese) economy. A diagram shows lines extending outward from Georgetown, Kentucky, where Toyotas are assembled, to twenty-two other sites, at distances ranging from 71 to 2,126 miles, where more than 440 U.S. suppliers manufacture parts for the Toyota. I consider that an achievement of design and management. But then I wonder whether we are about to confuse human life with an automobile. Will skilled managers take an egg from one source, a sperm from another, a gene from still another, a virus or enzyme and a retrovirus from other sources, and assemble an organism presumed to be superior to the works of nature? And in the attempt might we lose that relation to nature and God that Margaret Mead cherished (see Chapter 5)? I am not ready to prescribe limits to genetic adventures. I urge that everybody involved be sensitive to the issue.

A comic, yet serious dispute is symbolized in the contrast between two figures, Salvador Dali and Francis Crick. Dali's words, like his paintings, combine spoofery, exhibitionism, and seriousness, and all are evident in an interview with *Playboy* in 1964, where he claimed that one of his paintings was "an exact prophecy of the discovery of DNA." Praising the work of Watson and Crick as "the most important scientific event of our time," he said: "This is for me the real proof of the existence of God" (Dali 1964, 42, 48).

Dali gives no clue as to whether he knew that Francis Crick entered the field of molecular biology in order to confirm his own atheism (Judson 1997, 109). He could not know that Crick would one day, with tongue in cheek, use Dali's sentence as an epigraph for a book of his own, evidently because it represented exactly what Crick wanted to refute (Crick 1966). Crick here mounted a sustained argument against "vitalism," on the grounds that physics and chemistry could provide an adequate explanation for life.

Dali's statement, located among the pictorials and cartoons in *Playboy*, is not going to be a landmark in the exhaustive and exhausting literature on philosophical arguments for or against the existence of God. Nor is Crick's book, which misses the irony in the fact that, by its own logic, the uncomprehending physical-chemical molecular processes in one organism are trying to influence the uncomprehending physical-chemical molecular processes in other organisms, through the medium of words and logical reasons, to agreement. But if I find neither person persuasive, I find both to be representations of the differing ways in which different people respond to new knowledge that touches on the mystery of selfhood.

To join these two ways of seeing is a venture that extends far beyond the scope of this inquiry. It is far more than the universal field theory that Einstein sought so energetically and wistfully, because it includes not only the external world that Einstein investigated but also the minds and spirits of all investigators. The great thinkers of the ages have worked away at it. The new genetics touches on the fringes of it. But in the contradictions symbolized by Dali and Crick, it only echoes the controversies of the ages.

The cultural anthropologist Richard A. Schweder wisely observes: ". . . the world is incomplete if seen from any one point of view and incoherent if seen from all points of view at once" (Schweder 1994, 18). The best we can do now, perhaps forever, is state boldly the truths we see, recognize with modesty the truths that others see, and welcome whatever lucidity we can coopera-

Coda

tively find, without suppressing anything for the sake of a premature coherence.

Vincent Havel, longing for a "new world order" based on "universal respect for human rights," says "it will mean nothing as long as this imperative does not derive from the respect of the miracle of Being, the miracle of the universe, the miracle of nature, the miracle of our own existence" (Havel 1994).

Both ancient wisdom and the new genetics point up the perennial issue: we can see science as a continuous dissolution of miracle into a meaningless combination of chance and determinism; or we can see it as itself a miracle among the many miracles of cosmos and life. Science itself will not determine the total human response to science and its discoveries.

Reinhold Niebuhr, near the end of his illustrious career, wrote: "The mystery of human selfhood is only a degree beneath the mystery of God" (R. Niebuhr 1986, 256). This human self—in its biochemical processes, its exploits of thought, its hopes and visions—evokes the words of the Hebrew Psalmist (Psalm 139), words that may resonate even more deeply today than when they were first uttered:

> I will praise thee; for I am fearfully and wonderfully made: marvelous are thy works, and that my soul knoweth right well.

REFERENCES

This list includes only sources cited in the text, plus a dozen or so other works important to the argument of this book.

Anderson, W. French. 1987. Human gene therapy: scientific and ethical considerations. *Ethics, reproduction and genetic control.* Ed. Ruth F. Chadwick. London: Routledge, 1992.
Asimov, Isaac. 1962. *The genetic code.* New York: New American Library Signet Book.
Barbour, Ian. 1993. *Ethics in an age of technology.* The Gifford Lectures, Volume II. Harper San Francisco.
Barns, Ian. 1994. The human genome project and the self. *Soundings: an interdisciplinary journal* 77:99–128.
Benedict, Ruth and Gene Weltfish. 1943. *The races of mankind.* Included in *Race, science and politics* by Ruth Benedict. New York: Viking Press, 1950.
Bentham, Jeremy. 1789. *Introduction to the principles of morals and legislation.* Many editions.
Berger, Peter. 1963. *Invitation to sociology: a humanistic discipline.*

References

Garden City, NY: Doubleday & Company, Anchor Books.

Berger, Peter and Thomas Luckmann. 1967. *The social construction of reality*. Garden City, NY: Doubleday & Company, Anchor Books.

Birch, Charles and Paul Abrecht, eds. 1975. *Genetics and the quality of life*. Potts Point, New South Wales, Australia: Pergamon Press.

Boorstin, Daniel. 1975. Preface. *The timetables of history* by Bernard Grun. New York: Simon and Schuster.

Brando, Marlon with Robert Lindsey. 1994. *Brando: songs my mother taught me*. New York: Random House.

Broad, William and Nicholas Wade. 1982. *Betrayers of the truth: fraud and deceit in the halls of science*. New York: Simon and Schuster.

Bronowski, J. 1965. *Science and human values*. Revised edition. New York: Harper and Row Torchbook.

Carnegie Task Force on Meeting the Needs of Young Children. 1994. *Starting points: meeting the needs of our youngest children*. New York: Carnegie Corporation.

Caskey, C. Thomas. 1992. DNA-based medicine: prevention and therapy. *The code of codes: scientific and social issues in the human genome project*. Ed. Daniel J. Kevles and Leroy Hood. Cambridge, MA: Harvard University Press.

Catechism of the Catholic Church, The. 1994. Tr. copyrighted by the United States Catholic Conference. Mahwah, NJ: Paulist Press.

Cavalli-Sforza, Luca, Paolo Menozzi, and Albert Piazza. 1994. *The history and geography of human genes*. Princeton, NJ: Princeton University Press.

Chargaff, Erwin.
    1968. A quick climb up Mount Olympus. (A review of

References

Watson, *The double helix.*) *Science* 159:1448–49.
1974. Building the tower of Babel. *Nature* 248:776–79.
1978. *Heraclitean fire: sketches from a life before nature.* New York: Rockefeller University Press.
Cole-Turner, Ronald. 1993. *The new genesis: theology and the genetic revolution.* Louisville, KY: Westminster/John Knox Press.
Committee on Science and Technology, U.S. House of Representatives. 1982. *Hearings,* November 16, 17, 18 (No. 170). Washington: U.S. Government Printing Office.
Cook-Deegan, Robert Mullan. 1994. Germ-line therapy: keep the window open a crack. *Politics and the life sciences* 13(2) (August), 217–20.
Council for Responsible Genetics, 1992. *Position paper on human germ line manipulation.* Cambridge, MA: CRG.
Crick, Francis.
    1966. *Of molecules and men.* Seattle: University of Washington Press.
    1988. *What mad pursuit: a personal view of scientific discovery.* New York: Basic Books.
Dali, Salvador. 1964. Interview in *Playboy* 11, No. 7 (July).
Davis, Bernard. 1973. Threat and promise in genetic engineering. *Ethical issues in biology and medicine.* Ed. Preston Williams. Cambridge, MA: Schenkman Publishing Co.
Dawkins, Richard. 1989. *The selfish gene. New edition.* Oxford: Oxford University Press.
Descartes, René. 1641. *Meditations.* English translations in many editions.
Dobzhansky, Theodosius.
    1956. *The biological basis of human freedom.* New York: Columbia University Press.
    1967a. Changing man. *Science* 155 (1967): 411.

References

   1967b. *The biology of ultimate concern.* New York: New American Library.
Dubos, René. 1972. *A God within.* New York: Charles Scribner's Sons.
Duster, Troy. 1990. *Backdoor to eugenics.* New York: Routledge.
Eiseley, Loren.
   1956. *The immense journey.* New York: Random House Vintage Books.
   1960. *The firmament of time.* New York: Atheneum Publishers, pb.
ELSI bibliography: Ethical, legal and social implications of the human genome project, 1994 supplement. Washington: U.S. Department of Energy, Office of Energy Research.
Encyclopedia of bioethics. 1978. Warren T. Reich, editor in chief. 4 vols. New York: Free Press.
Fletcher, John C. and W. French Anderson. 1992. Germ-line therapy: a new stage of debate. *Law, medicine & health care* 20.
Fletcher, Joseph. 1974. *The ethics of genetic control: ending reproductive roulette.* Garden City, NY: Anchor Press/Doubleday.
Frankel, Mark. S. and Albert Teich, editors. 1994. *The genetic frontier: ethics, law, and policy.* Washington: American Association for the Advancement of Science.
Genewatch. A bulletin of the Council for Responsible Genetics.
Glass, H. Bentley. 1966. *Science and ethical values.* Chapel Hill, NC: University of North Carolina Press.
Graham, Loren R. 1981. *Between science and values.* New York: Columbia University Press.
Greenberg, Joel. 1981. B.F. Skinner now sees little hope for the world's salvation. *New York Times,* Sept. 15, C1.
Gustafson, James. 1973. Genetic engineering and the normative view of the human. *Ethical issues in biology and medicine.*

REFERENCES

Ed. Preston Williams. Cambridge, MA: Schenkman Publishing Co.

Haldane, J.B.S. 1963. Biological possibilities in the next ten thousand years. *Man and his future*. Ed. Gordon Wolstenholme. Boston: Little, Brown.

Hamer, Dean, Stella Hu, Victoria Magnuson, Nan Hu, and Angela Pattatucci. 1993. A linkage between DNA markers on the X chromosome and male sexual orientation. *Science* 261:321ff.

Hamer, Dean and Peter Copeland. 1994. *The search for the gay gene and the biology of behavior*. New York: Simon and Schuster.

Hamilton, Michael, ed. 1972. *The new genetics and the future of man*. Grand Rapids, MI: William B. Eerdmans Publishing Co.

Hastings Center Report. 1992, No. 4 (July–August). Special supplement: Genetic grammar: "Health," "illness," and the Human Genome Project.

Havel, Vaclav.
  1992. A dream for Czechoslovakia. *New York Review of Books*, June 25, p. 13.
  1994. The new measure of man. *New York Times*, July 8, 1994, A27 (op-ed).

Herodotus with an English translation by A.D. Godley. 1921, Vol. II. New York: G.P. Putman's Sons.

Herrnstein, Richard J. 1971. IQ. *Atlantic Monthly* 121 (Sept.):43–64.

Herrnstein, Richard J. and Charles Murray. 1994. *The bell curve: intelligence and class structure in American life*. New York: Free Press. (See also Murray and Herrnstein.)

Hoagland, Hudson and Ralph W. Burhoe, eds. 1962. *Evolution and man's progress*. New York: Columbia University Press.

Holland, Suzanne and Donna McKenzie. 1994. Genes, media and

policy. Unpublished paper, prepared for the Center for Theology and Natural Sciences, Berkeley, CA.

Howard, Ted and Jeremy Rifkin. 1977. *Who should play God?* New York: Dell Publishing Co.

Hubbard, Ruth and Elijah Wald. 1993. *Exploding the gene myth.* Boston: Beacon Press.

Human Genome News 6, No. 4, Nov. 1994.

James, William. 1896. The will to believe. In many collections of essays by James.

Judson, Horace Freeland. 1977. *The eighth day of creation.* New York: Simon and Schuster.

Kahn, Herman. 1971. Interviewed by G.R. Urban. *New York Times Magazine*, June 20.

Kevles, Daniel J. and Leroy Hood, eds. 1992. *The code of codes: scientific and social issues in the human genome project.* Cambridge, MA: Harvard University Press.

Kuhn, Thomas S. 1970. *The structure of scientific revolutions.* Second edition, enlarged. Chicago: University of Chicago Press.

Ledley, Frederick D. 1993. Genetics and the end of history. Unpublished paper presented to consultation on HGP, Center for Theology and the Natural Sciences, Berkeley, CA, Jan. 14–17.

Lee, Thomas F. 1991. *The human genome project: cracking the genetic code of life.* New York: Plenum Press.

Lewis, C.S. 1947. *The abolition of man.* New York: The Macmillan Co. The edition cited is the pb of 1965.

Longino, Helen E.

    1990. *Science as social knowledge: values and objectivity in scientific inquiry.* Princeton, NJ: Princeton University Press.

    1992. Essential tensions—phase two: feminist, philosophical, and social dimensions of science. *The social dimensions of science.* Ed. Ernan McMullin, Notre Dame, IN: University of Notre Dame Press.

Loury, Glenn. 1994. A political act. *New Republic* 211, No. 4 (Oct. 31).

Luria, Salvador. 1989. Letter to *Science* 246 (November): 873.

Maddux, John. 1994. Adventures in the germ line. *New York Times*, Dec. 11, 1994, Section 4, p. 15 (op-ed).

Mannheim, Karl. 1929, 1931. *Ideology and utopia.* Tr. Louis Wirth and Edward Shills. 1936. New York: Harcourt Brace and Co., pb ed.

Mead, Margaret. 1955. *Male and female: a study of the sexes in a changing world.* Second edition. New York: New American Library.

Mendel, Gregor. 1866. *Experiments in plant-hybridization.* Cambridge, MA: Harvard University Press, 1925.

Mill, John Stuart. 1861. *Utilitarianism.* Many editions.

Monod, Jacques. 1970. *Chance and necessity.* English tr., pb., New York: Vintage Books, 1972.

Morgenthau, Hans. 1972. *Science: servant or master?* New York: W.W. Norton & Co.

Muller, Herbert J. 1952. *The uses of the past.* New York: Oxford University Press. New York: New American Library Mentor Books, 1954. My citation is from the later pb. edition.

Muller, Herman J.

    1963. Genetic progress by voluntarily conducted germinal choice. *Man and his future.* Ed. Gordon Wolstenholme. Boston: Little, Brown.

    1967. What genetic course will man steer? *Proceedings of the third international congress of human genetics*, eds. James F. Crow and James V. Neel. Baltimore: Johns Hopkins Press.

REFERENCES

Murray, Charles. 1984. *Losing ground: American social policy.* New York: Basic Books.

Murray, Charles and Richard J. Herrnstein. 1994. Race, genes, and I.Q.:—an apologia. *New Republic* 211, No. 4 (Oct. 31), 27–37. (See also Herrnstein and Murray.)

Myrdal, Gunnar.
    1942. *An American dilemma.* Twentieth anniversary ed. 1962. New York; Harper & Row.
    1969. *Objectivity in social research.* New York: Random House.

Nature. News report, Sept. 29, 1994.

National Council of the Churches of Christ in the U.S.A. 1986. *Genetic science for human benefit.* A policy statement. New York: NCCUSA.

Nelkin, Dorothy and Laurence Tancredi. 1989. *Dangerous diagnostics: the social power of biological information.* New York: Harper Collins Basic Books.

Nelson, J. Robert. 1994. *On the new frontiers of genetics and religion.* Grand Rapids, MI: William B. Eerdman's Publishing Co.

New York Times, April 26, 1955, p. 17. Quoted in Spencer Weart, *Nuclear fear: a history of images.* Cambridge, MA: Harvard University Press, 1955.

Newsweek. 1993, July 26, p. 59. "Does DNA make some men gay? Science: the biology of destiny."

Nichols, Eve K. 1988. *Human gene therapy.* Cambridge, MA: Harvard University Press.

Niebuhr, H. Richard. 1989. *Faith on earth: an inquiry into the structure of human faith.* New Haven: Yale University Press.

Niebuhr, Reinhold.
    1932. *Moral man and immoral society.* New York: Charles Scribner's Sons.

References

1941. *The nature and destiny of man.* Vol. I, *Human nature.*

1986. A view of life from the sidelines. *The essential Reinhold Niebuhr: selected essays and addresses.* Ed. Robert McAfee Brown. New Haven, CT: Yale University Press. The specific essay was written in 1967, but first published posthumously in *The Christian Century* Dec. 19–26, 1984.

Nirenberg, Marshall. 1969. Will society be prepared? *Science* 157:633.

Okun, Arthur. 1975. *Equality and efficiency: the big tradeoff.* Washington: Brookings Institution.

Pascal, Blaise. 1670. *Pensées.* Tr. William Finlayson Trotter. New York: E.P. Dutton & Co., 1908.

Pope, Alexander. 1734. An essay on man. Many editions.

President's Commission for the Study of Ethical Problems in Medicine and Biomedical and Behavioral Research. 1982. *Splicing life: the social and ethical issues of genetic engineering with human beings.* Washington: U.S. Government Printing Office.

Ramsey, Paul.

    1970. *Fabricated man: the ethics of genetic control.* New Haven, CT: Yale University Press.

    1978. *Ethics at the edges of life.* New Haven, CT: Yale University Press.

Rifkin, Jeremy.

    1983a. *Algeny: a new word—a new world.* New York: Penguin Books.

    1983b. *Resolution.* Washington: Foundation on Economic Trends. Released June 8.

Rosenfeld, Albert. 1969. *The second genesis: the coming control of life.* New York: Random House, Vintage edition, 1975.

References

Santayana, George. 1955. *Scepticism and animal faith: introduction to a system of philosophy.* New York: Dover Publishers, 1955.

Segundo, Juan Luis. 1984. *Faith and ideologies.* Tr. John Drury. Maryknoll, NY: Orbis Books.

Shapiro, Robert. 1991. *The human blueprint: the race to unlock the secrets of our genetic script.* New York: St. Martin's Press.

Shinn, Roger L. 1991. Science, faith, and ideology in policy decisions. Ch. 12 of *Forced options: social decisions for the twenty-first century.* Third ed. with "Reconsiderations." Cleveland, OH: The Pilgrim Press.

Sinsheimer, Robert L. 1992. The prospect of designed genetic change. *Ethics, reproduction and genetic control.* Ed. Ruth F. Chadwick. London: Routledge.

Skinner, B.F.
    1971. *Beyond freedom and dignity.* New York: Alfred A. Knopf.
    1974. *About behaviorism.* New York: Alfred A. Knopf.

Suzuki, David and Peter Knudtson. 1990. *Genethics: the clash between the new genetics and human values.* Rev. and updated edition. Cambridge, MA: Harvard University Press.

Szasz, Thomas.
    1961. *The myth of mental illness.* Rev. edition. New York: Harper and Row.
    1994. *Cruel compassion.* New York: Wiley.

Teilhard de Chardin, Pierre. 1959. *The future of man.* English tr., New York: Harper & Row Publishers, 1964.

Thomas, Lewis. 1974. *The lives of a cell: notes of a biology watcher.* New York: Viking Press. Bantam pb, 1975. My citations are from the pb.

Time. 1993, July 26, front cover and pp. 36–39. Born gay?

REFERENCES

United Methodist Church. 1992. *Book of discipline of the United Methodist Church.* Nashville, TN: The United Methodist Publishing House.

Wade, Nicholas. 1979. *The ultimate experiment: man-made evolution.* New and expanded edition. New York: Walker and Co.

Watson, James D.

1968. *The double helix: a personal account of the discovery of the structure of DNA.* London: Weidenfeld and Nicolson. Penguin Books pb, 1970. My citations are from the pb.

1990. The human genome project: past, present, and future. *Science* 248:44–49.

1992. A personal view of the project. *The code of codes: scientific and social issues in the human genome project.* Eds. Daniel J. Kevles and Leroy Hood. Cambridge, MA: Harvard University Press.

Watson, John B. 1924. *Behaviorism.* New York: W.W. Norton.

Weber, Max.

1919a. Politics as a vocation. *From Max Weber: essays in sociology.* Eds. Hans H. Gerth and C. Wright Mills. New York: Oxford University Press. 1958.

1919b. Science as a vocation. From Max Weber: essays in *sociology,* op. cit.

Wexler, Nancy. 1992. Clairvoyance and caution: repercussions from the human genome project. *The code of codes: scientific and social issues in the human genome project.* Eds. Daniel J. Kevles and Leroy Hood. Cambridge, MA: Harvard University Press.

Whitehead, Alfred North. 1925. *Science and the modern world.* New York: Macmillan & Co.

REFERENCES

Wilson, Edward O.
   1978. *On human nature*. Cambridge, MA: Harvard University Press.
   1989. Threats to scientific diversity. *Scientific American* 261, No. 3 (Sept.) 108–16.

Wooldridge, Adrian. 1995. Bell curve liberals. *New Republic*, Feb. 27, 1995.

Wordsworth, William. 1978. Lines composed a few miles above Tintern Abbey. Many editions.

World Council of Churches.
   1981. *Manipulating life: ethical issues in genetic engineering*. Geneva: WCC.
   1989. *Biotechnology: its challenge to the churches and the world*. Report by WCC subunit on Church and Society. Geneva: WCC.

INDEX

Abortion, 36–37, 77
A,G,C,T, 26, 46, 51
AIDS, 32, 77, 83, 103, 136
Alcoholism, 33
Alzheimer's disease, 27, 33
American Association for the Advancement of Science (AAAS), 10, 11, 97
American Bar Association, 10
American Psychiatric Association, 101
Amniocentesis, 36
Amos, 75–76
Anderson, W. French, 41, 142–43
Anshen, Ruth Nanda, 10
Ants, 27
Arabs, 63
Aristotle, 70
Asilomar moratorium, 28
Asimov, Isaac, 141
Augustine of Hippo, 31

Babel, 143, 154
Babylonia, 63
Bacon, Francis, 119
Bacteria, 28, 38, 108
Barbour, Ian, 11, 72

Beethoven, Ludwig van, 109
Behaviorism, 48–49, 54
Bell curve, 55–67
Benedict, Ruth, 63–64
Bentham, Jeremy, 70, 135–36
Berger, Peter, 78–80, 100
Berra, Yogi, 46
Big Bang, 109
Bioethics, 75
Blood, 18
  Circulation of, 69–70, 107
Boorstin, Daniel, 62
Boston University School of Theology, 97
Brain, human, 64–65
Brando, Marlon, 128
Breast cancer, 43–44
Broad, William, 79
Buddha, Buddhism, 73, 109

Carter, President Jimmy, 73, 143
Caste, 50, 103
Carnegie Task Force, 64–65
Catholic Church, 126–27
Cavalli-Sforza, Luca, 61
Cells, somatic and germ cells, 23

[171]

INDEX

Center for Theology and the Natural Sciences, 10
Chance, 51
Chapman, Audrey R., 11,
Character, 117–19
Chargaff, Erwin, 10, 25, 27, 109
Chemical scissors, 28
Chimpanzees, 29, 108, 152
China, 63, 139
Christ, 109
Chorionic villus sampling, 36
Chromosomes, 23–28
Churchill, Winston, 46–47, 59, 83
Cognitive elite, 66
Cole, R. David, 10–11
Cole-Turner, Ronald, 15
Committee on Science and Technology, U.S. House of Representatives, 74
Confidentiality (see also, Privacy), 15
Congress, U.S., 13, 29, 95
Cook-Deegan, Robert M., 125, 218
Copernicus, Nicolaus, 19
Council for Responsible Genetics, 32, 125
Council of Europe, 124–25
Crick, Francis, 27, 29, 155–56
Culture (see also, Nature, Nurture), 60, 61, 99, 102–106, 114, 116, 120
Cystic fibrosis, 33, 77, 96, 133

Dali, Salvador, 155–56
Darwin, Charles, 18, 22
Davis, Bernard, 116, 127
Dawkins, Richard, 53
Declaration of Independence, 62, 85
Department of Defense, 83
Descartes, René, 50
Determinism (see also, Freedom), 49–55
Dewey, John, 70
Diabetes, 39, 127, 133
Dignity, human, 120, 136, 148, 153
Diversity, 112–15
DNA, 24–25, 26–29, 47–48, 70, 84, 88, 118, 127, 138
DNA fingerprinting, 15, 151–52

Recombinant, 27–28
Dobzhansky, Theodosius, 10, 55, 133, 141–42
Double helix, 27, 29–30, 111
Double standard in ethics, 17–18
Down syndrome, 37
Dualism, 50
Dubos, René, 55
Dwarfism, 38

Economic issues, 15, 39, 151
Economics and research, 31–32
Egypt, 63
Einstein, Albert, 50, 59, 60, 61, 109, 156
Eiseley, Loren, 117
ELSI, 30–32, 35, 74
Encyclopedia of Bioethics, 75
Episcopal Church, 126
Eskimos, 113
Ethics (see also, ELSI, Values), 31, 48
  Double Standard, 17–18
  Exploratory, 150–51
  Ethical inquiry, 31
  Prescriptions, 15–16, 89, 107
  Qualified autonomy of, 73, 81, 89, 147–48
  Theory and method, 13–14, 93
  Tragic element in, 137
Eugenics, 99–100, 128, 137–41
  "Hard," 140
  "Soft," 140–41

Faith, 72–73, 87, 89, 106, 149–50
Facts, 79, 87
Fertilization *in vitro*, 41
Fletcher, John C., 142–43
Fletcher, Joseph, 109
Flynn effect, 65
Freedom, 45–67, 92
Freedom-in-community, 110–12, 149

Galileo, Galilei, 19, 22, 77
Galton, Charles, 138
Gandhi, Mohandas, 107, 118

INDEX

Gender
   And abortion, 36–37, 77
   And double standard, 17–18
   And ideology, 86–88
   And science, 81
Gene, genes, 25–27, 29, 35, 39–41, 42–44, 48, 54–55, 108
   "Selfish," 53
   Gene-splicing, 28, 30, 37–38
Genesis, Book of, 21
   Genesis 30:37–43, 18
Genghis Khan, 106
Genome, defined, 29
Genotype, 20–21
Germ cells, 23
German Parliament, 125
Germline therapy, 123–45
Glass, H. Bentley, 10, 133
Gore, Al, 74
Graduate Record Examination, 60
Greece, 63
Greenberg, Joel, 49
Gustafson, James, 96–97

Haldane, J.B.S., 118
Hamer, Dean, 43
Harvey, William, 70
Havel, Vaclav, 111–12, 157
Health, norms of, 95–121
Heisenberg, Werner, 50, 109
Hermeneutics of suspicion, 87
Herodotus, 98–100
Herrnstein, Richard J., 55–67, 129
Hinduism, 50, 63
Hitler, Adolf, 59, 106
Holland, Suzanne, 42
Homosexuality, 42–43, 101
Hormones
   For cows, 38
   Human growth, 38–39
Howard, Ted, 143
Hubbard, Ruth, 119, 141
Human Genome News, 31

Human Genome Organization (HUGO), 29
Human Genome Project (HGP), 13, 29–32, 70, 73–75, 79, 81–84, 93, 95, 150
Huntington's disease, 33, 35, 39–40, 77, 96, 130, 148

Incest, 18
Ideology, 47, 85–89, 106, 134–35, 150
Image of God, 120, 136
Imagination, 115–17
India, 63
   Indians, Callation, 98
Informed consent, 111, 136–37
Instinct, 116
Insulin, 37–38, 39
Intelligence (IQ), 55–67, 88, 115
Interferon, 38
Isaiah, 106
Islam, 73

Jacob (biblical), 18
James, William, 90, 149–50
John Paul II, Pope, 144
Jonas, Hans, 97
Jordan, Michael, 61
Joyce, James, 9
Judson, Horace Freeland, 15, 155

Kahn, Herman, 30
Kant, Immanuel, 31, 70
Keillor, Garrison, 39
Kevles, Daniel J., 100
King, Martin Luther Jr., 47, 118
King, Mary-Claire, 43
Kuhn, Thomas S., 30

Ledley, Fred, 47
Lee, Thomas F., 23, 48
Lenin, V.I., 119
Lewis, C.S., 141
Loury, Glenn, 62
Love, 117

Index

Luckmann, Thomas, 79, 100
Luria, Salvador, 140–41
Lutheran denominations, 126
Lysenko, Trofim D., 77

Maddux. John, 145
Malaria, 33
Manipulation, 111
Mannheim, Karl, 86
Mapping of genome, 29
Marx, Karl, 86
  Marxist ideology, 139
Masai, 104–105
Maya, 63
McKenzie, Donna, 42
Mead, Margaret, 10, 109–10, 144, 155
Medicaid, 39
Mendel, Gregor, 19, 22–24, 33, 35, 59
Mesopotamia, 63
Mill, John Stuart, 136
Monod, Jacques, 27, 51, 53
Monogenic ailments, 33, 40
Mouse, mice, 26–27, 108, 130–31, 152
Muller, Herbert J., 99
Muller, Herman, 25, 27, 91, 118–19
Murray, Charles, 55–67, 129
Muscular dystrophy, 33
Mutation, 24, 114, 115
Myrdal, Gunnar, 72–73, 79
Mystery, 52, 156–57

National Council of Churches of Christ in the U.S.A., 73, 125, 126, 143, 145
National Institutes of Health, 28, 40, 83
Natural selection, 18, 105
Nature (and nurture), 45–67, 102–106, 108
Nazis, Nazism, 99–100, 138–139, 140
Newsweek, 43
Newton, Isaac, 19, 60, 61
Niebuhr, H. Richard, 45, 52
Niebuhr, Reinhold, 82, 118–19, 157
Nirenberg, Marshall, 138
Norms of humanity, 95–121

Skewed, 98–101
Illusions of perfection, 101–102
Nucleotides, 26, 29, 111, 120
Nurture (see Nature)

Obesity, 33
Okun, Arthur, 59

Pascal, Blaise, 99, 153
Patents, 15, 151
Peters, Ted, 10
Phenotype, 20–21
Picasso, Pablo, 61, 113–14
Pindar, 98
PKU (phenylketonuria), 34
Plato, Platonism, 8, 31, 50, 70, 73, 98–99, 100, 119
Plants, genetically modified, 40
"Playing God," 143–45, 154–55
Pluralism, cultural and religious, 72, 90–93, 98–101, 150
Polemics, 16
Politics, 81–85, 91, 147
Polygenic ailments, 33–40
Pope, Alexander, 153–54
Positivism, 73, 78
Prenatal diagnosis, 36–37
Presbyterian Church, 126
President's Commission, 73, 143
Privacy (see also, Confidentiality), 151
Probability, 21
Prometheus, 143
Psalm 100, 144
Psalm 139, 157
Public policy (see also, Politics), 69, 72–73, 150
Pygmies, 104–105

Race, 37, 61–67, 86, 103–104
Reason, 115–17
Relativism, cultural and ethical, 98–101
Religions, 36, 50, 52, 73–74, 106, 137
Religious communities, 73–74, 89–91
Rembrandt, H. van Rijn, 61

[174]

Index

Restrictive enzymes, 28
Retrovirus, 28
Rifkin, Jeremy, 125, 143
Risk, 38, 39–40, 108, 131–32, 134, 135–37
Romans, 63
Rosenfeld, Albert, 15
Roosevelt, Franklin D., 59
Ruth, Babe, 61

Saddam Hussein, 107
Santayana, George, 149
Sartre, Jean-Paul, 52
Schweder, Richard A., 156
Science, 30, 74–81, 84, 147
 And ethics, 14, 69–93, 129–35, 142
 Qualified autonomy of, 77, 130, 147–48
 "Value-free," 78–81, 148–49
Scott, J.B., 117
Sequencing of genes, 29
Serendipity, 80
Servetus, Michael, 70
Shakespeare, William, 18
Shapiro, Robert, 30, 129
Siblings, 24
Sickle-cell anemia, 33–34, 133
Sinsheimer, Robert L., 128
Skinner, B.F., 49
Sophists, 99
Sperm banks, 119
Spinoza, Benedict de, 31, 70
Stalin, Joseph, 59
Stoicism, 50
Switzerland, 125
Synagogue Council of North America, 73, 143
Szasz, Thomas, 100

Tay-Sachs disease, 33–35, 77, 96, 133
Teilhard de Chardin, Pierre, 128
Ten Commandments, 82
Theology, 90–93, 144

Therapy, genetic, 15
 Somatic, 39–41, 152
 Germline, 41–42, 123–45, 152
 And social control, 100–101, 137–41
Theresa, Mother, 107
Third (Two-Thirds) World, 139
Thomas, Lewis, 23, 127
Time, 43
Tobacco, 77–78
Toyota automobiles, 155
Tutsi, 104

UNESCO, 127–28
United Church of Canada, 126
United Church of Christ, 126
United Methodist Church, 125, 126
United States Catholic Council, 73, 143
Utilitarianism, 73, 135–36

Values, 71–74, 84, 89, 106, 134, 135–43, 147
 Qualified autonomy of, 73, 81, 89, 147–49
Vectors, 27, 40

Wade, Nicholas, 15
Wald, Elijah, 119, 141
Watson, James, 27, 29–30, 47–48, 79, 140
Watson, John B., 48–49
Watutsi, 104
Weber, Max, 78–80, 90
Weltfish, Gene, 63
Wexler, Nancy, 35, 80–81
Whitehead, Alfred North, 16, 121
Wilkins, Maurice, 29
Wilson, Edward O., 113, 128–29, 133
Wooldridge, Adrian, 60
Wordsworth, William, 154
World Council of Churches, 125, 126, 137–38, 145

Zero, 63

COLOPHON

Roger L. Shinn is the Reinhold Niebuhr professor emeritus of social ethics at New York's Union Theological Seminary. The present book comes out of thirty years of participation in North American and international projects seeking to define the social, ethical, and religious issues involved in the new genetics. The efforts involved cooperation of biologists, lawyers, social scientists, philosophers, and theologians. With Margaret Mead, he co-chaired a Task Force on The Future of Man in a World of Science-Based Technology. A former president of the American Theological Society and the Society of Christian Ethics, he has lectured in many universities and professional societies on the subject of this book.

The text was set in Caslon, a typeface designed by William Caslon I (1692-1766). This face designed in 1725 has gone through many incarnations. It was the mainstay of British printers for over one hundred years and remains very popular today. The version used here is Adobe Caslon. The display face is Melior.

Composed by Alabama Book Composition, Deatsville, Alabama.

The book was printed by Thomson-Shore, Dexter, Michigan on acid free paper.